# 3D Printing of Concrete

*Series Editor*
*Gilles Pijaudier-Cabot*

# 3D Printing of Concrete

*State of the Art and Challenges of the
Digital Construction Revolution*

*Edited by*

# Arnaud Perrot

WILEY

First published 2019 in Great Britain and the United States by ISTE Ltd and John Wiley & Sons, Inc.

Apart from any fair dealing for the purposes of research or private study, or criticism or review, as permitted under the Copyright, Designs and Patents Act 1988, this publication may only be reproduced, stored or transmitted, in any form or by any means, with the prior permission in writing of the publishers, or in the case of reprographic reproduction in accordance with the terms and licenses issued by the CLA. Enquiries concerning reproduction outside these terms should be sent to the publishers at the undermentioned address:

ISTE Ltd
27-37 St George's Road
London SW19 4EU
UK

www.iste.co.uk

John Wiley & Sons, Inc.
111 River Street
Hoboken, NJ 07030
USA

www.wiley.com

© ISTE Ltd 2019

The rights of Arnaud Perrot to be identified as the author of this work have been asserted by him in accordance with the Copyright, Designs and Patents Act 1988.

Library of Congress Control Number: 2019930612

British Library Cataloguing-in-Publication Data
A CIP record for this book is available from the British Library
ISBN 978-1-78630-341-7

# Contents

**Introduction** . . . . . . . . . . . . . . . . . . . . . . . . . . . . . . . . . ix

Arnaud PERROT

**Chapter 1. 3D Printing in Concrete: General
Considerations and Technologies** . . . . . . . . . . . . . . . . . . . 1

Arnaud PERROT and Sofiane AMZIANE

   1.1. Introduction . . . . . . . . . . . . . . . . . . . . . . . . . . . . . . 1
   1.2. General considerations for 3D printing and
   additive fabrication . . . . . . . . . . . . . . . . . . . . . . . . . . . . 2
      1.2.1. What is 3D printing? . . . . . . . . . . . . . . . . . . . . . . 2
      1.2.2. Towards the 3D printing of cement-based materials . . . . . 7
   1.3. The digital and additive fabrication of cement materials . . . . . 7
      1.3.1. Introduction . . . . . . . . . . . . . . . . . . . . . . . . . . . 7
      1.3.2. Printed methods using extrusion and deposition . . . . . . . 9
      1.3.3. Methods of printing by injection into a
      particle bed . . . . . . . . . . . . . . . . . . . . . . . . . . . . . . . 22
      1.3.4. Alternative printing methods . . . . . . . . . . . . . . . . . . 25
   1.4. A classification of 3D printing methods for concrete . . . . . . . 29
      1.4.1. Philosophy . . . . . . . . . . . . . . . . . . . . . . . . . . . . 29
      1.4.2. Classification parameters . . . . . . . . . . . . . . . . . . . . 30
      1.4.3. Example of classification . . . . . . . . . . . . . . . . . . . . 33
   1.5. References . . . . . . . . . . . . . . . . . . . . . . . . . . . . . . . 35

## Chapter 2. 3D Printing in Concrete: Techniques for Extrusion/Casting ... 41
Arnaud PERROT and Damien RANGEARD

2.1. Introduction... 41
2.2. Breakdown of the process into stages ... 43
2.3. Behavior during the fresh state and the printing stage ... 46
    2.3.1. Rheology of cement-based materials ... 46
    2.3.2. Pumping ... 52
    2.3.3. Extrusion ... 54
    2.3.4. Stability of an elemental layer during deposition ... 56
    2.3.5. Overall stability of the printed structure in a wet state... 58
2.4. Other problems occurring during concrete extrusion printing ... 62
    2.4.1. Elastic deformation and accuracy of the deposition ... 62
    2.4.2. Shrinkage and cracking during drying... 63
    2.4.3. Bonding between layers – weakness at the interface between layers... 65
    2.4.4. Concept of time windows ... 66
2.5. Conclusion ... 67
2.6. References ... 68

## Chapter 3. 3D Printing by Selective Binding in a Particle Bed: Principles and Challenges ... 73
Alexandre PIERRE and Arnaud PERROT

3.1. Introduction... 73
3.2. Classification of selective printing processes and strategies... 75
    3.2.1. Selective cement activation ... 77
    3.2.2. Selective paste intrusion ... 80
    3.2.3. Injection of the binder ... 82
3.3. State of the art of selective printing and major achievements... 82
3.4. Scientific challenges... 84
    3.4.1. Selective cement activation and the effect of water penetration ... 84
    3.4.2. Selective intrusion and penetration by cement paste... 89
    3.4.3. Towards modeling in 3D ... 94

| | |
|---|---|
| 3.5. Conclusion. | 96 |
| 3.6. References. | 96 |

## Chapter 4. Mechanical Behavior of 3D Printed Cement Materials . . . . . . . . . . . . . . . . . . . . . . . . . . . . 101

Mohammed SONEBI, Sofiane AMZIANE and Arnaud PERROT

| | |
|---|---|
| 4.1. Introduction. | 101 |
| 4.2. Mechanical performance of the cement materials printed using the extrusion/deposition method . | 102 |
| 4.2.1. Effect of extrusion on the mechanical characteristics of cement-based composites. | 103 |
| 4.2.2. Mechanical behavior of 3D printed cement materials. | 105 |
| 4.3 Effects of the additive fabrication method on the mechanical behavior of cement-based materials . | 116 |
| 4.3.1. Printed concrete = anisotropic stratified materials: possible causes . | 116 |
| 4.3.2. Effects of the printing process parameters on the mechanical properties . | 116 |
| 4.4. Mechanical behavior obtained with other methods of 3D printing of cement-based materials . | 119 |
| 4.4.1. Production using robotic sliding castings ("Smart Dynamic Casting") . | 119 |
| 4.4.2. Printing using the method of injection into a particle bed . | 119 |
| 4.5. Conclusion. | 120 |
| 4.6. References. | 121 |

## Chapter 5. 3D Printing with Concrete: Impact and Designs of Structures . . . . . . . . . . . . . . . . . . . . 125

Arnaud PERROT and Damien RANGEARD

| | |
|---|---|
| 5.1. Introduction. | 125 |
| 5.2. Freedom of forms: architectural liberation and topological optimization. | 126 |
| 5.2.1. 3D printing with concrete: a boon for architects? . | 126 |
| 5.2.2. Towards the creation of structures with optimized shapes? . | 128 |
| 5.2.3. Could 3D concrete printers go through a transition similar to the transition from black and white to color? . | 130 |

5.3. Design of structures: reinforcement strategies
and design codes . . . . . . . . . . . . . . . . . . . . . . . . . . . . . 131
   5.3.1. The use of fibers . . . . . . . . . . . . . . . . . . . . . . . . 132
   5.3.2. External reinforcements . . . . . . . . . . . . . . . . . . . . 133
   5.3.3. Steel wire placed within the extruded material. . . . . . . . 133
   5.3.4. Dedicated spaces acting as lost formworks. . . . . . . . . . 134
   5.3.5. Wrapping of reinforcement elements set
   in place beforehand. . . . . . . . . . . . . . . . . . . . . . . . . . . 135
   5.3.6. Towards a specific design code? . . . . . . . . . . . . . . . 136
5.4. Impacts of 3D printing . . . . . . . . . . . . . . . . . . . . . . . . 136
   5.4.1. Environmental impact . . . . . . . . . . . . . . . . . . . . . 136
   5.4.2. Societal impact . . . . . . . . . . . . . . . . . . . . . . . . . 138
   5.4.3. Economic impact . . . . . . . . . . . . . . . . . . . . . . . . 139
5.5. Conclusion . . . . . . . . . . . . . . . . . . . . . . . . . . . . . . . 140
5.6. References . . . . . . . . . . . . . . . . . . . . . . . . . . . . . . . 141

**List of Authors** . . . . . . . . . . . . . . . . . . . . . . . . . . . . . . . 145

**Index** . . . . . . . . . . . . . . . . . . . . . . . . . . . . . . . . . . . . . 147

# Introduction

## I.1. Context of the book

The ability to generate a three-dimensional (3D) object from a single image would seem like an idea pulled from a work of fantasy or science fiction. Nevertheless, starting from the mid-1980s, with the first patents on additive manufacturing and 3D printing, this future possibility has become a reality. Initially limited to polymers, additive manufacturing has now expanded to an ever-increasing number of materials [HUL 86, AND 84]. In the 2000s, the development of fused deposition modeling, or the rapid prototyping through the deposition of polymer strands, led to a rapid democratization of the process and gave the general public a glimpse of the ample possibilities offered by 3D printing, in terms of economic and industrial development. In addition, this technology is perfectly suited to the societal environmental issues currently faced: in that it first enables us to save materials for the manufacturing of parts with complex geometry, and second, to consider the "on demand" manufacturing of spare parts.

Naturally, the possibility of transferring these technologies into the field of construction, including concrete, was initially studied by Pegna in 1997 [PEG 97] and then by Professor B. Khoshevnis of the University of Southern California in the first half of the 2000s [KHO 04]. At the same time, the computer-aided design of structures

---

Introduction written by Arnaud PERROT.

has undergone significant advances with the introduction of the first digital models (building information modeling – BIM) [HEG 01].

As a result, traditional building methods may now find themselves overthrown by the Third Industrial Revolution and by the introduction of computerization and digital technologies. In this context, the design and monitoring of projects have already been influenced by the use of BIM: the creation of complete digital models of buildings compelled further action to be taken in the period leading up to the execution of a construction project, allowing for further steps towards both an optimization of the execution methods and optimal construction quality (referred to as lean building).

Nevertheless, the use of digital technologies in production methods is still in its infancy (prototypes, feasibility and reliability in the laboratory). However, the opportunity to take advantage of the complete digitization of construction projects, starting from the time they are designed, enables us to envision the automation of construction methods and to make greater progress towards the objectives of lean building.

Thus, the application of additive manufacturing methods, originally developed for plastics, to concrete is now the subject of numerous academic studies and private initiatives around the world. As a result, the number of initiatives and projects related to the 3D printing of concrete has grown exponentially since 2015. For example, Figure I.1 shows the growth of the number of publications on the topic of concrete additive manufacturing in the 10 most influential scientific journals in the field of civil engineering (source: Google Scholar, date: July 1, 2018). The late 2000s and early 2010s saw the publication of pioneering works by Professor Khoshnevis of the University of Southern California and the team of Professor Buswell of Loughborough University in England [KHO 04, KHO 06, BUS 07, LE 12, LEA 12]. Since 2016, there has been an explosion in the number of publications that show the current nature of this research area and the need for knowledge related to the area of concrete 3D printing in construction works. While there were four such publications in 2016, 16 were found in 2017 and 33 in the first six months of 2018.

The motivation for these studies can be found in:

– the economic advantage offered by 3D printing, which could offer the potential of avoiding the use of concrete forms, representing up to 50% of the cost of cast concrete;

– unprecedented freedom for architects in the shapes they can create;

– a reduction in environmental impacts – the ability to place the material exclusively where it is needed (known as the concept of topological optimization);

– improvements in working conditions – elimination of heavy handling tasks and the vibration of concrete.

These initial works have made it possible to validate the technical feasibility of the process of 3D printing with concrete, and small-scale demonstrators have been carried out around the world (individual homes, walkways).

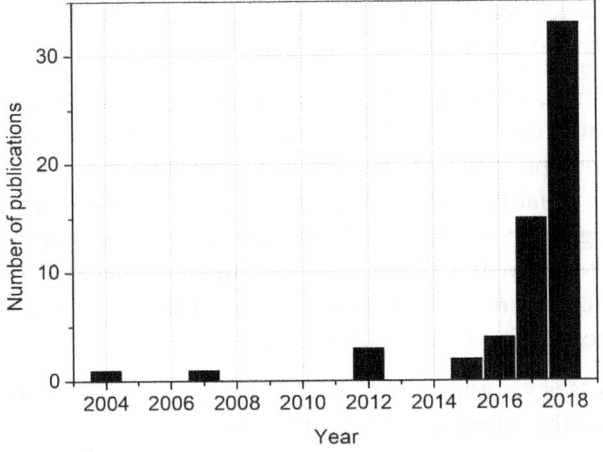

**Figure I.1.** *Number of publications in the top 10 most influential science journals in civil engineering (source: Google Scholar, as of July 1, 2018) over the past 15 years*

The market for printed concrete is now worth nearly €30 million, and is now growing at a rate of 15% per year. The exponential growth in the number of projects has made it possible to imagine an extremely rapid increase in the revenue of this market.

As a result, the application of printed concrete in structural material no longer looks like a utopian vision, and it is now important to lay the foundations for these new manufacturing techniques by carrying out a comprehensive compendium of the knowledge and technologies developed in the field.

## I.2. Current research topics and scientific challenges

Currently, the research topics related to 3D printing are:

– Shift towards a 100% digital construction industry

The design of a construction project involves the production of a digital mock-up, which leads to both the anticipation of construction problems and the optimization of the interactions between the different professional occupations and stages of construction. This anticipation tool enables us to improve the quality of the constructions and to optimize the methods of execution.

In addition, these digital models make up a raw material that can be implemented using robots and automation schemes, allowing construction projects to be carried out faster, more accurately and more reliably. The construction of concrete structures using 3D printing fits perfectly within this framework. An efficient transfer interface between the digital model and the trajectory of the robot placing the concrete is expected to be achieved while taking into account the configuration of the construction site and its constraints.

– Processes: optimization and mastery of the rheology of the concrete for the purpose of using it for printing

To be able to be printed, the concrete must not only be fluid enough to be transferred (pumpable concrete) but also rigid enough to hold up under its own weight once extruded, without deforming. Similarly, it must "quickly" bear the weight of the layers placed on top of it. The competition between the rate of mechanical structural build-up and that of the elevation of the printed structure is therefore a critical parameter to be controlled in order to ensure that the process is carried out smoothly. This involves controlling not only the behavior of concrete in the state of being freshly placed but also the changes it undergoes over time. It is also important to gain control over the additives used, which allows the material to be put in place "on demand" (many processes involve the addition of an accelerator in the nozzle of the printer). Works on the mechanical behavior of concrete at a very early age will be necessary to describe the behavior of the material up to the end of its placement, and thus to allow its transition with the initial rheological behavior. It is important not only to work on experimental methods for describing the evolution of the rheological behavior (shear yield stress) of concrete over time in a simple and reliable way, but also to be able to control and follow the process inline.

It should also be noted that other innovative processes, such as deposition on a support, injection into aggregate beds or through meshes or porous structures and "intelligent" sliding forms, are conceptually similar technologies that will need to be studied.

– Structural design of printed structures

- Characterizing and reinforcing an anisotropic material

Printed concrete set in layers may exhibit anisotropic behavior, induced by its layered structure, which should be qualified. The interface between layers, depending on the process, remains a sensitive area, which may represent the mechanically weak points of the structure. It will therefore be necessary to establish a study methodology to characterize the complex and anisotropic behavior of this type of material.

Furthermore, printed concrete, similar to poured concrete, has a tensile weakness that will have to be compensated for by the addition of steel reinforcements, following the same principles of the reinforcement traditionally used for poured concrete.

Several approaches are currently being tested: the addition of fibers within the layers, the casting of steel bars in dedicated spaces and the addition of a metal wire to the concrete. The effectiveness of these reinforcement methods remains to be evaluated, and strategies for their scaling are to be developed accordingly.

- Topological optimization

Additive manufacturing technologies enable us to envision a new level of freedom in structural design inspired by nature (biomimicry) that optimizes the management of resources by using materials only where they are mechanically necessary. This leads to the possibility of lean manufacturing and mechanically optimized structures that go beyond the design codes of traditional concrete structures. It is therefore necessary to provide a normative framework for the design of structures printed from concrete.

– Sustainability and environmental benefits

An important first step will be to assess the impact of the stratification induced by the implementation of printing techniques on the durability of the printed concrete. Subsequently, the use of 3D printing makes it possible to envision a saving in materials (and the transporting of materials) by means of topological optimization. On the contrary, the rheological properties required for the process require the extensive use of additives. A comparative analysis of the environmental impacts of the two modes of implementation will enable us to measure the environmental benefits achieved by the use of 3D printing, which makes it possible to print a structure with optimal volumes of materials compared with a casted concrete structure.

## I.3. Structure of the book

In order to help give readers an idea of the state of the art in the area of digital concrete production and to give them a frame of reference with regard to the current issues listed above, we have organized this book in the following way: after this introduction, all the current technical solutions will be presented. Then, the main families of concrete printers will be presented, as well as the machines that best represent each of the categories. A principle for classifying the printing systems on the basis of the relevant literature will also be presented.

Next, the aspects of the process related to materials will be presented, first by addressing the aspects related to the technique of printing via the successive extrusion/layering of materials. This technique is directly inspired by the fusion/placement technique used by polymer 3D printers that are becoming more democratic today. Second, the material constraints of the process of particle-based 3D printing, that is, injection into a particle bed, will be addressed.

Then, the mechanical behavior of mortar and concrete generated through printing will be addressed, focusing on the peculiar aspects of these materials in comparison with conventional cast concrete.

Finally, the potential impacts of the methods of digital production on structural design and the economics of construction, and the environmental impacts of the sector will be addressed. The reinforcement systems to be put in place, ensuring equivalent mechanical characteristics, will also be described, with the aim of presenting the strategies for the design of printed concrete structures.

This scientific work structure will make it possible to assess the current state of the techniques of additive manufacturing as applied to cement-based materials, addressing both the scientific and technological aspects.

## I.4. References

[AND 84] ANDRE J.-C., LE MEHAUTE A., DE WITTE O., "Dispositif pour réaliser un modèle de pièce industrielle", *FR Patent* 2,567,668, 1984.

[BUS 07] BUSWELL R.A., SOAR R.C., GIBB A.G. *et al.*, "Freeform construction: Mega-scale rapid manufacturing for construction", *Automation in Construction*, vol. 16, no. 2, pp. 224–231, 2007.

[HEG 01] HEGAZY T., ZANELDIN E., GRIERSON D., "Improving design coordination for building projects. I: Information model", *Journal of Construction Engineering and Management*, vol. 127, no. 4, pp. 322–329, 2001.

[HUL 86] HULL C.W., "Apparatus for production of three-dimensional objects by stereolithography", *Google Patents*, 1986.

[KHO 04] KHOSHNEVIS B., "Automated construction by contour crafting–related robotics and information technologies", *Best ISARC 2002*, vol. 13, no. 1, pp. 5–19, January 2004.

[KHO 06] KHOSHNEVIS B., HWANG D., YAO K.-T. et al., "Mega-scale fabrication by contour crafting", *International Journal of Industrial and Systems Engineering*, vol. 1, no. 3, pp. 301–320, 2006.

[LE 12] LE T.T., AUSTIN S.A., LIM S. et al., "Mix design and fresh properties for high-performance printing concrete", *Materials and Structures*, vol. 45, no. 8, pp. 1221–1232, 2012.

[LEA 12] LEACH N., CARLSON A., KHOSHNEVIS B. et al., "Robotic construction by contour crafting: The case of lunar construction", *International Journal of Architectural Computing*, vol. 10, no. 3, pp. 423–438, 2012.

[PEG 97] PEGNA J., "Exploratory investigation of solid freeform construction", *Automation in Construction*, vol. 5, no. 5, pp. 427–437, 1997.

# 1

# 3D Printing in Concrete: General Considerations and Technologies

## 1.1. Introduction

The application of 3D printing and the introduction of digital technologies into the concrete construction sector are the result of the appropriation and repurposing of technology initially developed for other materials, such as resins and plastics.

This would be a good time to introduce the need to improve performance on the construction site.

The construction industry accounts for 13% of industrial expenditures worldwide, but only 1% of annual growth in productivity. It is a sector that has a long tradition of low productivity and stunted technological capabilities, in comparison with the technical developments in other industries that have largely automated and digitized their manufacturing processes. The reasons for this conservative outlook are numerous and often stem from the uniqueness of each building, which, in practice, does not allow processes to be reproduced and, to a certain extent, limits the technical framework on the site. The low productivity brought about by methods that are a burden to implement results in high production costs, which make it difficult for companies to develop. For instance,

---

Chapter written by Arnaud PERROT and Sofiane AMZIANE.

operations using concrete castings account for 50–75% of the cost of the structural work, while concrete and steel materials account for only a small percentage of manufacturing costs.

Automation on construction sites is one of the methods suggested by the study published in 2017 by McKinsey [MK 17]. The introduction of a disruptive technology such as 3D printing would bring the construction industry into the digital realm, leading to a rise in the technical levels of workers and expanding the possibilities for building design and architecture.

In order to better understand 3D printing methods, first, we will provide an overview of 3D printing in general, as well as the different methods initially developed. Next, we will present the different methods used for concrete, inspired by the initial techniques used for plastics, ceramics and metals. Finally, in the last section, we will attempt to classify the concrete printing methods proposed in the literature [DUB 17].

## 1.2. General considerations for 3D printing and additive fabrication

### 1.2.1. *What is 3D printing?*

There is no universally recognized definition of 3D printing.

Many different definitions and terms are used or combined to describe 3D printing, such as additive fabrication, computer-aided or digital fabrication, or rapid prototyping. In a general sense, we can define it as "a process of assembling materials to make objects or structures from a 3D data model, usually layer by layer, as opposed to traditional methods of production by subtraction" [GAR 11].

3D printing is a booming technology that is widely regarded as an industrial revolution. Thus, it is important to become familiar with some of the basic notions about 3D printing in general, the processes it utilizes, the way it works and its limitations.

Methods of design using 3D printing make it possible to shorten design and development times, improve collaboration between different parties and therefore help solve problems that arise between the worlds of engineering and design.

Although it has been used for a long time, especially for the production of prototypes, the term "3D printing" is now generally used to describe the additive fabrication method, regardless of the specific technology, material or application involved.

### 1.2.1.1. *Main concepts*

In additive fabrication, we can identify different manufacturing philosophies, with different methods and elements to use. All of these involve a series of steps in their design and rapid prototyping that form a specific cycle.

Generally, they consist of the choice of a material used as the basis for the construction of the model, as well as a sophisticated computer system that controls the processes of adding materials, such as depositing, sintering, injection, melting, etc.

Thus, the process begins with the creation of a 3D model in CAD (computer-aided design) format, which must be converted into STL (stereolithography) format. This format is processed by specific software programs, typically used in additive fabrication, which cuts the object into "slices" to create a new file containing the information for each layer.

This cycle of design and manufacturing can therefore be summarized in the following way:

– digital modeling: creation of a 3D model, a true digital representation of what you would like to create. To do this, we will use CAD software. Additive fabrication is inextricably linked with the digital design of parts and structures;

– exporting: generation of a file in a dedicated format (STL) containing all the geometrical information needed to represent the digital model;

– stratification: conversion of the digital model into a list of commands that the 3D printer can understand and carry out;

– connection: establishment of a list of instructions given to the printer through a data sharing system;

– printing: preparation of the 3D printer and beginning of the printing itself;

– finishing: final stage consisting of the extraction of temporary parts and the addition of non-printable finishing parts.

### 1.2.1.2. *Classification of printing technologies*

Existing 3D printing technologies can be classified in many ways. In 2009, the American Society for Testing and Materials (ASTM) and the Society of Manufacturing Engineers (SME) created the committee "F42 for Additive Manufacturing Technology" [AST 12, MON 15]. A classification of ASTM's activities groups these technologies into six categories [DIL 17].

### 1.2.1.2.1. Material extrusion

The material is extruded through a nozzle to form multilayered patterns. This method is commonly used for plastic materials, which is known as the fused deposition modeling (FDM) technique. It is also the most popular and least expensive among professional polymer printers.

In this case, the most commonly used materials are polycarbonates, although it is possible to add different materials such as wood, minerals or rubbers to a polymeric matrix.

### 1.2.1.2.2. Photopolymerization

A deposit of liquid photopolymer resin is hardened by selective exposure to light (via a laser or projector), which initiates the polymerization and solidifies the exposed areas.

This technique is associated with stereolithography (SLA). This system uses a beam of light to selectively solidify resins or suspensions sensitive to light [LIA 96]. It is one of the most widely used methods in additive manufacturing.

### 1.2.1.2.3. Fusing and direct application of energy

A layer of powder is first deposited. Depending on the part to be printed, a part of the powder is then agglomerated by melting it together using a heat source. The operation uses a tool such as a laser or an electron beam. The powder surrounding the consolidated part that has not melted yet acts as a supporting material.

The printed form is obtained by repeating this operation of successive deposits of layers of powder (known as beds of particles). This method can be used for plastics or metals.

It is interesting to note that this technique can also be used with a direct application of energy on the material, but only in the volume of the structure to be printed. In this case, no material is used in excess. This technique is known as direct application of energy.

### 1.2.1.2.4. Injection of binder: binder jetting

Liquid binders are applied on the layers of powder grain material to build the various parts. The binders can be both organic and inorganic.

Three-dimensional printing (3DP) uses this process, which combines granulates and binders [GRA 97]. Each layer is created by a thin layer spread with a roller, causing the granules to bond into a solid by adding a binder using a printing head. Then, the printing carriage moves downward and creates the next layer of aggregate. This process is used to manufacture many components made from metals, ceramics and polymers.

### 1.2.1.2.5. Injection of materials: jetting

Drops of material are deposited layer-by-layer for the manufacturing of parts. We can use photosensitive resin jets and UV rays, or jets that deposit melted materials that solidify at room temperature.

This method is used by multi-jet modeling (MJM) printing, which deposits small drops of polymeric materials onto a platform with thin sheets. The material of the final model is solidified using UV lamps, while the supporting gel is removed with water jets. This method enables the creation of details with a very precise surface finish [BHA 16].

Another system derived from this technology is the Thermojet, which deposits drops of a melted ceramic material which then cool and solidify, forming a solid material.

### 1.2.1.2.6. Lamination of sheets

This is a method for cutting and layering. The fabricating of the laminate is done via a system in which each part is built sequentially by sheets of material. This process consists of thermal bonding of a series of parts with uniform thickness, cut using a laser. The system includes a 2D printer located on a work table that moves along the vertical axis.

The 2D printer creates layers from one sheet of material, each of which joins the next with thermo-adhesives located on one side of the sheets. A tool bonds the sheets by applying force and heat to each of the layers.

The layers are finally superimposed, and the geometric definition of the object will be defined by the thickness of the object.

Each technique is shown in Figure 1.1.

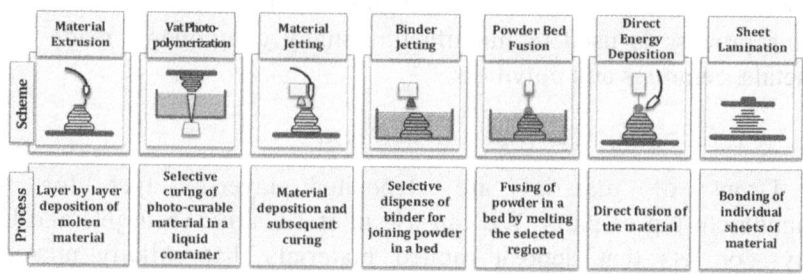

**Figure 1.1.** *Visualization of different 3D printing methods (based on [DIL 17])*

## 1.2.2. *Towards the 3D printing of cement-based materials*

Once the different techniques for 3D printing are defined, it becomes clear that the process requires a fast, even nearly instantaneous solidification of the materials in order to achieve a rapid printing rate. This meant that the thermo-dependent nature of the mechanical behavior of materials such as resins, polymers or even metals would need to be utilized to solidify and aggregate the materials and obtain the part to be printed. Thus, the application of external energy sources (laser, UV rays, etc.) for phase changing or softening the material is necessary in traditional 3D printing processes.

It is clear that these techniques cannot be applied to cement-based materials, and the rigidity of the material cannot be increased simply by relying on the structure of the cement during the dormant period (a type of behavior referred to as "thixotropic" [ROU 12, ROU 06, PER 15, LEC 17]), nor by relying on the phenomenon of cement hydration.

## 1.3. The digital and additive fabrication of cement materials

### 1.3.1. *Introduction*

In this section, we will list all the classifications of the methods used for the 3D printing of cement-based materials. For each of these approaches, we will attempt to present the different philosophies, and provide the existing and/or imaginable parameters and variants.

The first method, which is the most common and the most extensively developed at this time, is the extrusion/deposition method. This technique will be the starting point for our overview of 3D printing techniques. As we will see, this technique makes it possible to manufacture structures at different scales, from small elements to entire buildings.

Next, the method using injection into a bed of particles is used in the field of cement-based materials, and has made it possible to obtain small elements [PIE 18] as well as elements of structures, such as a recently opened pedestrian bridge in Madrid [ALL 16].

Currently, these two methods are most commonly used, and most frequently described in the literature. Finally, we will take a look at alternative methods, such as the method of projection onto mobile supports, the method of controlled slip-forming processes and the method of injection into a mesh screen produced by robots (permeable formwork).

It should be noted that all of these methods are different from each other. The goal here is to list the existing techniques and highlight their differences and the areas to which they can be applied.

As an example, Lim et al. proposed a classification [LIM 12] with three additive D-Shape manufacturing processes (printing by injection into a particle bed), Contour Crafting (production of walls using the extrusion/deposition technique) and Concrete Printing (deposition of multiple mortar filaments in order to make complex structures) (Figure 1.2).

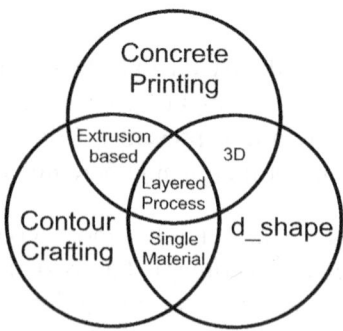

**Figure 1.2.** *Classification of printing methods proposed by Lim et al. [LIM 12]*

Given the recent developments made, we propose the following extension of this classification:

– extrusion/Contour Crafting;

– printing by injection into a particle bed/D-Shape;

– controlled slipforms: "Smart Dynamic Casting";

– injection into printed wire mesh permeable formworks "Mesh Mould";

– shot concrete onto a mobile device.

## 1.3.2. *Printed methods using extrusion and deposition*

### 1.3.2.1. *Ensure the mixing, transport and deposition*

During the printing of concrete or cement materials, it is first necessary to be able to transport the concrete via a container, a mixer or even a concrete mixer truck to the printing head.

The mixing of materials is done in a relatively conventional way, with the mixing of powders and granulates into a homogeneous mixture, followed by the addition of the liquid phases in general until the desired rheology is obtained. Superplasticizers and accelerators can be added during extrusion, as we will see below. Recently, it has been shown that the energy used in mixing can positively influence the rate of the structure of cement-based materials [VAN 19].

A pump can be used to deliver the material to the extrusion head. In this case, the power of the pump and the technology it uses will have to be chosen based on the rheology of the concrete and the length of the flexible hoses required for its delivery. This pumping step could limit the initial consistency of the material to be printed (essentially, the more fluid a concrete is, the easier it is to pump).

Another solution is to locate a tank for the concrete on site, using a feeder screw that leads directly to the printing head [ASP 18]. However, in this case, the amount of material available is limited.

Therefore, it is necessary to include a re-feeding system for large-scale printing tasks.

The two solutions can be mixed and a rotating screw can be incorporated into the printing head in order to provide not only partial relief for the pump but also additional energy to allow for the reduction in the diameter at the nozzle of the extruder [KEA 13] (Figure 1.3).

At the level of the print head, several adjustments and tools can be added to improve the printing process. These can include the addition of accelerator agents to allow a rapid rise in the resistance of the first printed layers that can be found in many processes. This principle is shown in Figure 1.4. Thus, the material is fluid in the pumping hoses, which facilitates the pumping stage, but quickly becomes rigid once it is deposited, which ensures the stability of the structure during printing.

Other accessories for the nozzle have been tested or are under development. Examples of these include a smoother/scraper system to remove irregularities in the printed structure and limit the roughness of the layers, or a total or partial sealing system to reduce and control the section of the material leaving the extruder.

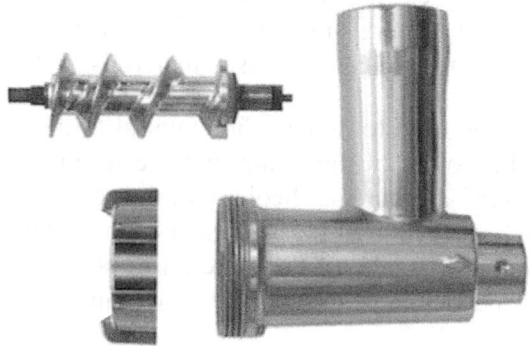

**Figure 1.3.** *Example of a printing head with a feeder screw [KEA 13]*

Finally, it is important to note that the cross-section of the deposited material can have several forms. In the literature, examples of somewhat narrow forms of rectangular or cylindrical cross-sections can be found. In all cases, once it is deposited, the cross-section of the material must be controlled to achieve the desired shape of the printed structure.

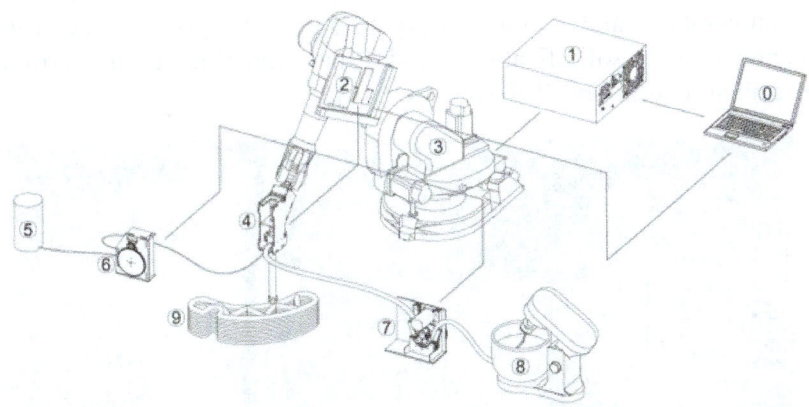

**Figure 1.4.** *Schematic representation of the printing of building components using an articulated arm (based on [GOS 16]). 0. Control system; 1. robot control center; 2. printing control; 3. robotic arm; 4. printing head; 5. accelerator agent; 6. pump for accelerator-type additive; 7. mortar pump; 8. mortar mixer; 9. object printed in 3D. For a color version of the figures in this chapter see www.iste.co.uk/perrot/3dprinting.zip*

### 1.3.2.2. *Presentation of the extrusion/deposition method*

The extrusion/deposition method for the printing of cement-based materials is the first method to be developed in the field of concrete construction, thanks to the work of Professor Khoshnevis and his team [KHO 04, KHO 06] in the early 1990s. It is now the most widely studied and most developed printing technique for industrial applications. In fact, many initiatives in France (XtreeE, Construction-3D, INHOVA), Germany (Project ConPrin3D), the Netherlands (Project Milestone, TU Eindhoven), China (WinSun company), Russia (Apis Cor), United States (Contour Crafting) and others rely on this method of printing.

The principle of printing by extrusion and deposition is based on the method of 3D printing via the deposition of melted polymers (FDM). It consists of extruding layers of cement-based material to create a component or structure following the digital model built using design software (Figure 1.5).

Unlike polymers, the behavior of concrete has very little sensitivity to temperature, and it is necessary to adjust the printing speed and geometry to the solidification speed of the cement material controlled by cement hydration [PER 16, WAN 16].

**Figure 1.5.** *Deposition of layers of mortar during printing via extrusion/deposition of cement-based materials*

### 1.3.2.3. *Different robots for different scales of execution*

Extrusion/deposition printing technologies have been and continue to be developed to print elements and structures of different sizes and under different conditions. Depending on the size of the elements to be printed (elements of a building, small constructions of structures such as single-family homes, or multi-level buildings), different robotic technology will be used to move the printing head.

### 1.3.2.3.1. Printing construction elements

For the printing of construction elements, several robotic technologies are available. The most commonly used technology is a six-axis robotic arm that allows work to be carried out with a volume from 1 up to several cubic meters (Figure 1.6). In this case, the process involves the inclusion of a system for routing the materials (such as a pump) and a depositing system (such as a printing or extrusion head) in order to be able to deposit the material in a system initially intended to follow programmed trajectories. Relatively complex systems are added to adjust the advancement rate of the robot to the output flow rate of the material.

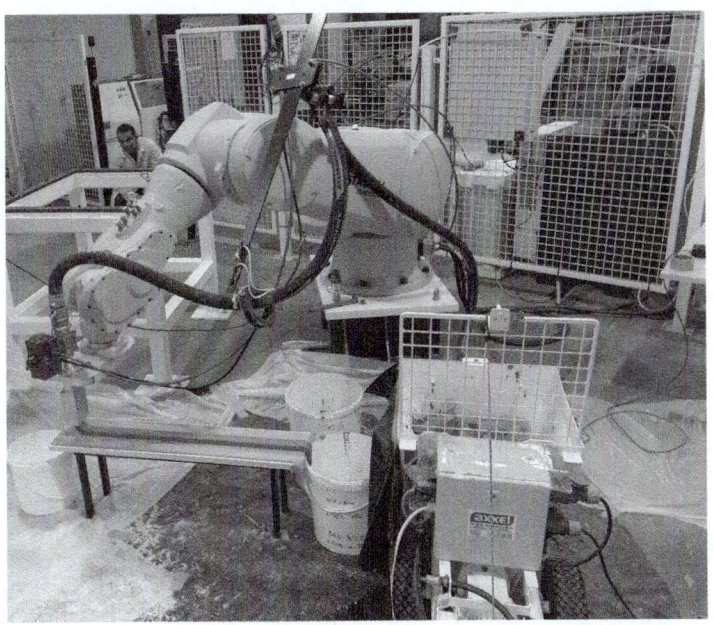

**Figure 1.6.** *A robotic arm used for the 3D printing of cement-based materials*

This technology is used by companies such as XtreeE [France] for the production of their printed objects, as well as in many academic works.

Technology based on the controls for bridge cranes is also a common solution for the printing of objects and structural elements made of cement-based materials. For example, the team led by Dr Buswell at Loughborough University [LIM 12, BUS 07], the first printings made by Professor Khoshnevis, and more recently, the work done at the University of Eindhoven [BOS 16, BOS 18] use this technique (Figure 1.7). The fundamental idea on which this technique is based is similar to the operation of conventional paper printers, using a crane that can be positioned over the bridge surface, adding to that the height control of the crane relative to the level of the object that has already been printed. This technique is also used for polymer printers that print via the deposition of melted materials.

**Figure 1.7.** *3D printing of concrete parts using a printer with bridge cranes [LIM 12]*

Finally, the team of Professor Asprone of Federico II University of Naples uses a delta robot developed by the company WASP to produce prefabricated components of beams that are then assembled after printing (Figure 1.8) [ASP 18]. The use of the delta robot enables the rapid positioning of the robot. However, the simple use of this type of robot does not allow the orientation of the printing head to be changed, as is the case with the six-axis robot (as well as bridge crane printers).

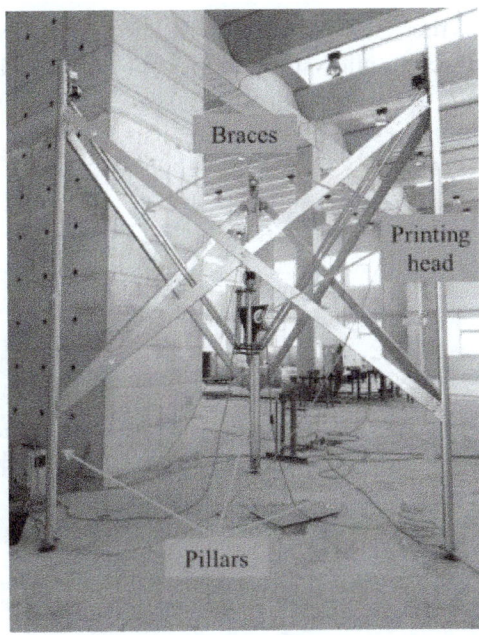

**Figure 1.8.** *A delta robot used for the 3D printing of mortar-based elements of beams [ASP 18]*

It is important to note that when producing building components with these technologies, the distance to be covered by the robot to place a layer is small. As a result, the fabricating and lifting of printed objects can be done very quickly. Thus, it is recommended for the cement material to have stiffening and structural build-up rates that are in accordance with the load related to the weight of the printed structure.

### 1.3.2.3.2. Printing small constructions

The development of 3D printing is also a solution that offers a fast and economically advantageous means for the production of single-family housing [POU18]. With this goal in mind, several technological approaches have been developed for the on-site construction of small homes. Most of them are based on the scaling up of solutions developed for producing small components.

For example, the use of fixed robotic arms has been proposed by different companies to allow the creation of curved walls and vertical support elements of single-family houses. This solution has been offered by companies such as the Russian firm Apis Cor or Construction-3D (Figure 1.9).

**Figure 1.9.** *A fixed robotic arm used for the construction of vertical carriers for individual houses by the company Apis Cor [API 18]*

In addition to these, the Spanish startup Be More 3D, in collaboration with the University of Valencia, has developed a 3D printer capable of printing houses with an area of 24 m² in 12 hours. This time can be reduced to 8 hours with printing at maximum speed. Finally, tests were also made for using a robotic arm, but in this case a mobile one, that is, one that is mounted on a mobile vehicle. The Yhnova project in Nantes has made it possible to print a lost formwork made from polyurethane foam, which acts as thermal insulating material and filled with self-compacting concrete. These walls are used for building single-family homes in Nantes [POU 18]. It also represents a global first, in the sense that the house is inhabited and considered to comply with safety and comfort standards, demonstrating the full extent of the mechanical robustness of the process (Figure 1.10).

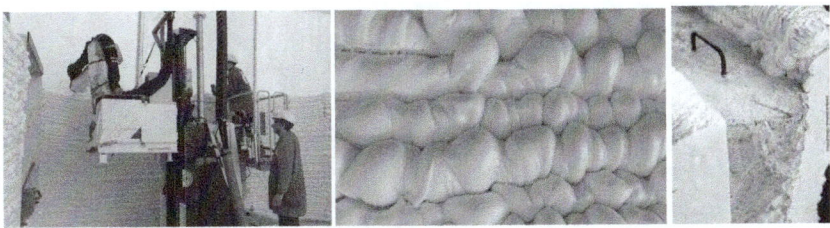

**Figure 1.10.** *A mobile robotic arm used for the implementation of walls made of formworks printed from expanded foam filled with self-compacting concrete for a single-family house in Nantes in 2017 [POU 18]*

The firm Cazza Construction (United States) has also developed a mobile robot of this type [CAZ 17] (Figure 1.11).

**Figure 1.11.** *A mobile robotic arm used for the construction of vertical walls for individual houses by the company Cazza Construction [CAZ 17]*

We might therefore imagine implementing a team of multiple robots to increase and optimize production rates.

Finally, the installation of overhead bridges makes it possible to envisage the construction of single-family homes as imagined by Professor Khoshnevis using the Contour Crafting technique (Figure 1.12).

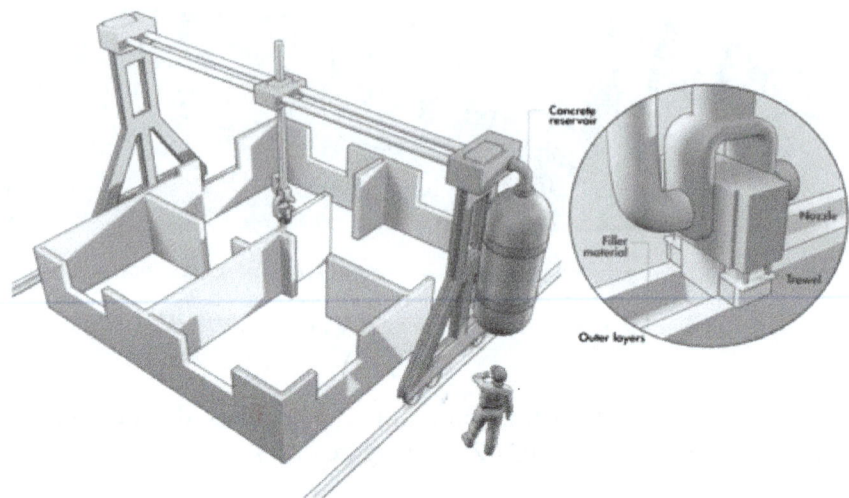

**Figure 1.12.** *An automated overhead bridge on rollers for the construction of a single-family home (based on [ZHAN 13])*

### 1.3.2.3.3 Shift towards printing multi-story buildings

The scaling up to the construction of multi-story buildings is currently under development. This will require the development and adaptation of lifting and scaling equipment to program the trajectories over several blocks of ten-meter spans.

Several solutions are currently under development, such as the use of robots with cables (Figures 1.13 and 1.14), fixed cranes or crane trucks (Figure 1.15).

The concept of the use of cable robots was conceived more than ten years ago [BOS 05, BOS 07]. In 2017, this technique was used for the manufacturing of raw earth structures [IZA 17]. The installation of structures for robot cables appears to be relatively easy, and allows large surface areas to be covered. The main problem faced by this method is the ability to generate the paths taken by the robot without any interaction between the printed structure and the path of the printing head.

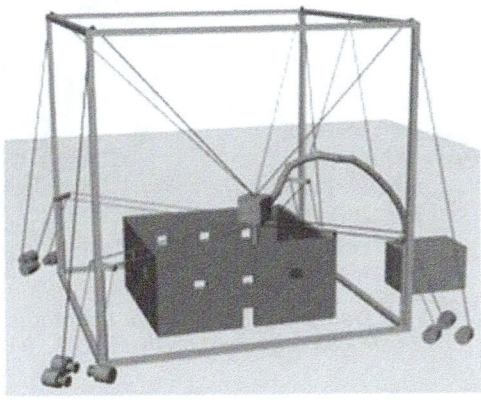

**Figure 1.13.** *Principle of the use of cable robots in concrete 3D printing (based on [BOS 05])*

**Figure 1.14.** *Use of a cable robot for the production of structures from mud-brick (according to [IZA 17])*

The use of cranes and crane trucks also seems to offer a promising solution, in which they are converted for use as concrete printers capable of carrying out a volume of work on the scale of the space taken up by a multi-story building. These solutions are currently in the conceptual stage [LIA 14] or in the development stage, such as the ConPrin3D project developed by the Dresden Technical University [NER 16a, NER 16b] (Figure 1.15).

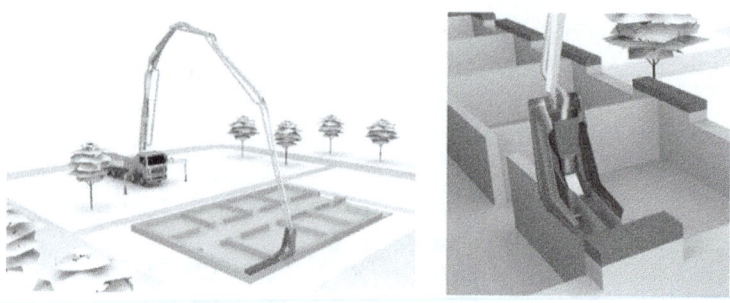

**Figure 1.15.** *Use of a robot truck for the production of structures (based on [NER 17])*

For the transition to this scale of printing, a significant amount of work is required on controlling the positioning, and on how to take environmental conditions (wind, rain) into account. To do this, it will be essential to install gyroscopes and techniques for compensation. This is the subject of the patent developed by the team formed by Amziane, Godi and Phelipé at Clermont Auvergne University, which has installed gyroscopic systems to stabilize the printer attached to a crane. This design is capable of printing a multi-story building, since the crane covers an area of action on the scale of the building (Figure 1.16).

**Figure 1.16.** *Use of a crane equipped with a gyroscope and tracking system, and a print head to print the concrete structures*

### 1.3.2.4. Difficulties and limitations

Several difficulties remain to be overcome in order to truly achieve the democratization of 3D printing by extrusion/deposition on building sites.

In reality, the multi-layered aesthetic appearance that is a drawback of this process does not always produce the most positive effects. On the contrary, the mechanics of the joints at the interfaces between layers is a problem which should be sufficiently analyzed or even eliminated. In addition, we must take care to ensure that this interface does not provide a quick path of entry for agents that are aggressive towards cement-based materials.

Moreover, in the case of conventional concrete construction, it will be necessary to find ways to strengthen the concrete by means of reinforcement structures that can provide good mechanical properties to the materials in traction. One preliminary solution may be the production of shapes (such as arches), which can limit tensile stress. External fibers or reinforcements can also be used, followed by the pouring of reinforced concrete in the dedicated voids left in the printed structure.

These issues are also linked to a lack of normative regulation in the area of printed concrete. Commissions and working groups are currently being set up, particularly in scientific societies such as the RILEM (French acronym for the International Union of Laboratories and Experts in Construction Materials, Systems and Structures) or the ACI (American Concrete Institute).

Furthermore, within the context of a construction site, the accuracy of the control will be paramount, and the demands related to the external environment will have to be taken into account in order to achieve an acceptable level of accuracy in the construction. It is an issue of the robustness of the processes that are presented here. At this time, printed constructions have been carried out under controlled conditions, and it will be necessary to be able to implement robust solutions to compensate for the effects of wind speeds, hygrometry and temperature.

Furthermore, the solutions for printing cement-based materials by extrusion/deposition limit the freedom of architects. Effectively, the transition from a vertical structure to a horizontal structure can only take place very gradually, because the cement-based materials in their fresh state cannot be deposited with too much of a cantilever without becoming deformed. The use of supports, or the printing of temporary supports with another material, might solve this limitation.

### 1.3.3. *Methods of printing by injection into a particle bed*

#### 1.3.3.1. *Presentation of this method*

The method of printing via an injection of fluids into a bed of particles is a printing method based on those developed for polymers and metals. The goal of this is to spread out a bed of particles of a given thickness, and then to make injections of a fluid in the locations dictated by the numerical model allowing the grains to be agglomerated.

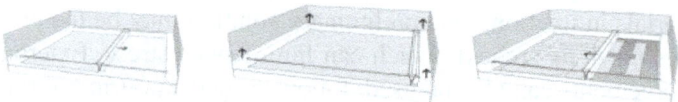

**Figure 1.17.** *Execution sequence of 3D printing by injection into a particle bed: 1) spread of a layer of particles; 2) lowering the tray; 3) localized injection of the fluid [LOW 18]*

Two primary methods have been developed for cement-based materials. The first is to inject a very fluid cement paste into a bed of aggregates (sand and/or gravel) [PIE 18, LOW 18]. The second method consists of injecting water (including some adjuvants if necessary) into a granular bed containing the cement-based binder and the aggregates [LOW 18, SHA 17, WEG 16]. The second method has also been used for Geopolymers [XIA 16].

The pioneer of this method in the field of building materials is Enrico Dini, creator of the D-Shape method [LOW 18, CES 14].

These methods theoretically allow complex structures to be obtained with contour precision in the order of magnitude of the diameter of the largest grain found in the granular bed. This enables an excellent printing resolution. Figure 1.18 shows the surface quality obtained as a function of the size of the granulates used.

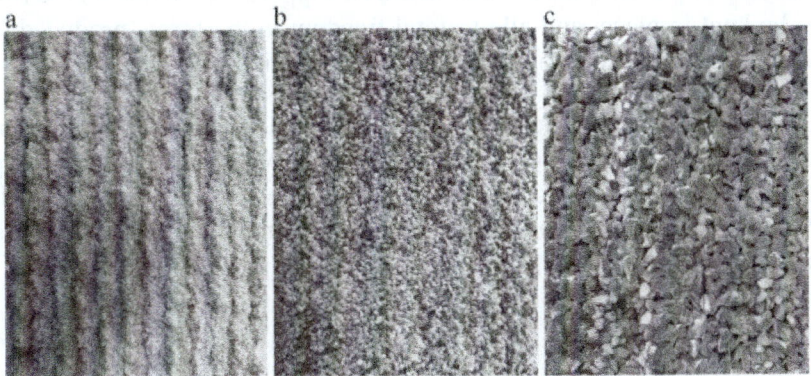

**Figure 1.18.** *Surface condition obtained as a function of a particle bed with granulates of the size: a) 0.2 mm; b) 2 mm; c) 4 mm [LOW 18]*

### 1.3.3.2. Different scales of execution

The method of injection into a particle bed can be used at the same time to make small components (Figure 1.19) [PIE 18]. For this type of execution, a small printer holding in a volume of one cubic meter is sufficient (Figure 1.19).

**Figure 1.19.** *Example of small components and the printer used [PIE 18]*

It is interesting to note that the object printed in Figure 1.19 contains a sudden change between a vertical structure and a horizontal structure. The bed of particles acts in this case as a support, which allows the fluid to be kept in place, and allows total freedom in the forms produced (as is not the case for printing by extrusion/deposition).

The works of Enrico Dini have also made it possible to create objects and structures on larger scales [LOW 18, 44]. The D-Shape printer that he designed (Figure 1.20) has made it possible to build a pedestrian walkway and a structure several meters high, both at the same time (Figure 1.21).

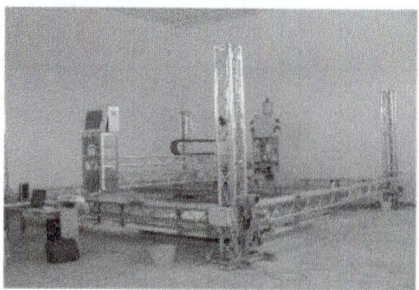

**Figure 1.20.** *A D-Shape printer and an injection system [LOW 18]*

**Figure 1.21.** *Example of objects created with the D-Shape printer [ALL 16, 44]*

### 1.3.3.3. Difficulties and limitations

One of the major difficulties in the process of printing based on particle bed injection is choosing the properties of the fluid (rheology, surface tension) according to the physical parameters of the granular bed (pore size, total porosity). If the fluid is not selected on this basis, and it penetrates the layer incompletely, the mechanical behavior of the printed concrete will be poor. This fine level of adjustment is difficult to understand [PIE 18, LOW 18] and requires a fine level of analysis, which will be dealt with in Chapter 3.

The removal of particles that are not connected via the binder is also a necessary step, which is incorporated once the printing has finished. This step can take a long time and can be difficult for complex geometries.

The shift to a larger scale, the scale of multi-story structures, is also made more difficult by the management and placement of large quantities of particles.

### 1.3.4. Alternative printing methods

Several original methods, which take into account the specific characteristics of fresh cement materials, have also been developed recently and tested at the laboratory level. The main original methods created for concrete are presented in the following sections.

#### 1.3.4.1. Controlled slip forming: "Smart Dynamic Casting"

One of the first methods for the digital construction of concrete structures is based on the slip-forming method, which has been used for decades for the construction of large vertical elements (water towers, bridge piers, etc.). This technique, known as Smart Dynamic Casting, was developed at ETH Zurich in the early 2010s [WAN 16, LLO 15, SHA 13]. It consists of the robotizing of the process of the slip forming, which takes advantage of the transition of the behaviors between fresh, malleable concrete and the material once it has hardened. Smart Dynamic Casting enables the production of geometrically complex, vertical concrete structures without the need for custom-made formworks. To do this, it uses a form with a flexible or rigid outlet nozzle (which can potentially include a variable section die), attached to a six-axis robotic arm which forms the concrete at the time of its exit (Figure 1.22). The robotic arm allows a dynamic casting of the concrete.

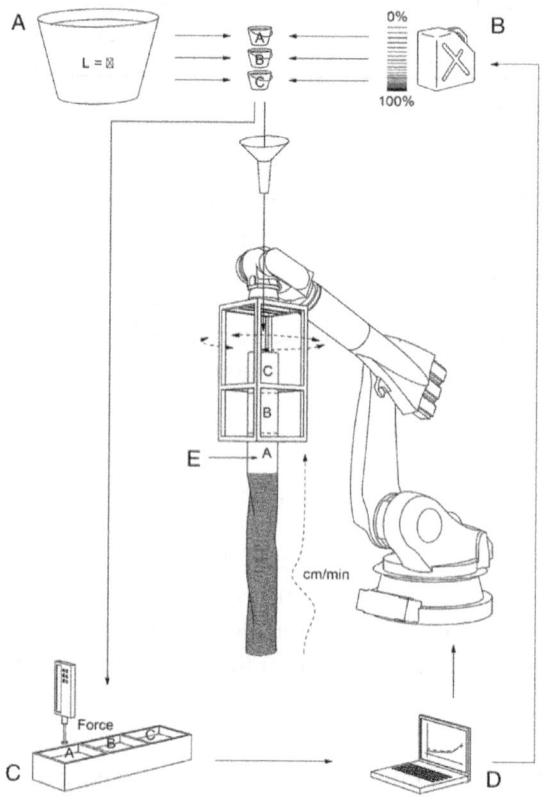

**Figure 1.22.** *Principle of the Smart Dynamic Casting process [LLO 15]: A) a large volume of self-compacting concrete is prepared with a setting retardant; B) addition of the accelerator with the placing of the castings; E) setting the form by placement in the castings; Steps C) and D) correspond to an in-line control of the rheology progression, which enables the rate of lifting the forms to be controlled*

In-line rheology control makes it possible to determine the speed at which the forms must be lifted. If the forms are lifted very quickly, the concrete will be too fluid to support its weight. On the contrary, if the lifting is too slow, the stress from friction in the contact between the forms and concrete will become very strong, and therefore the printed structure may break [SHA 13]. Furthermore, the in-line modification of the rheology by the addition of an accelerator or a hydration retardant is possible during the preparation of the material, its introduction or its passage through the nozzle.

The Smart Dynamic Casting method can therefore be used to produce vertical elements with complex and optimized geometries (Figure 1.23). The technique is currently being improved to produce angular, slender and curved walls [GRA 19].

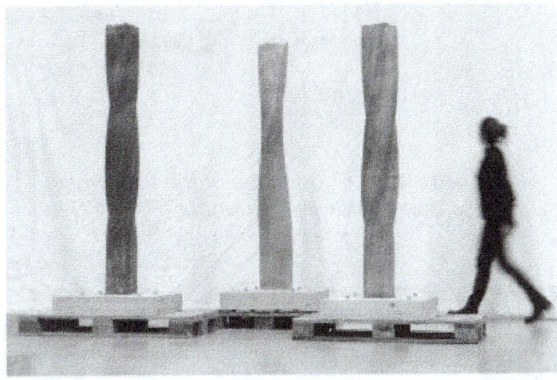

**Figure 1.23.** *Example of elements produced using the Smart Dynamic Casting technique [LLO 15]*

### 1.3.4.2. *Injection into printed wire mesh castings (Mesh Mould)*

Another technique developed at ETH Zurich is the use of a welding robot (an articulated arm mounted on a track) to produce a steel rebars net as a permeable formwork for concrete that is cast in it later [HAC 14, HAC 15]. The size of the formwork meshes must be set in accordance with the rheology of the material, so that there will be a sufficient flow of concrete to coat the mesh, but not too much to be able to hold the form made by the mesh (Figure 1.24). Once the concrete has been poured, the surface must then be finished in a tapering step, which can also be done by robots (Figure 1.25).

This technique is interesting, because it enables the production of steel reinforced wall elements with a behavior similar to that of conventional reinforced concrete.

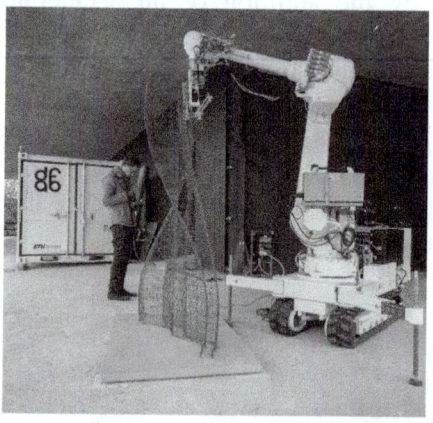

**Figure 1.24.** *Production of the formwork screen by the welding robot [HAC 17]*

**Figure 1.25.** *Structure obtained after the casting of the concrete and scraping to achieve the finish [HAC 17]*

### 1.3.4.3. *Shot concrete onto a mobile device*

The last alternative method presented in this chapter is the method of projection onto mobile supports [LIN 18a, LIN 18b, NEU 16]. This method is based on the methods using robotic technology to shot concrete into a tunnel or onto a vertical surface [WIE 17].

The concrete could be shot directly or onto a mobile support that temporarily supports the concrete while it is produced. The advantage of this method is that the technologies for transporting and accelerating the setting of the concrete that has been shot are well known, and could improve production rates.

This method could also facilitate the transition from a horizontal to a vertical structure through the firmness of projected mixtures [LIN 18a, LIN 18b].

## 1.4. A classification of 3D printing methods for concrete

### 1.4.1. *Philosophy*

The wide variety of concrete printing processes presented in this chapter makes it necessary for them to be classified so that the methods for manufacturing can be obtained in accordance with the complexity, size and, more generally, the specifications of the objects to be created.

To facilitate this choice, a classification of concrete printing processes has been proposed by Duballet *et al.* [DUB 17]. Based on the parameters of the classification related to both the process and the part to be printed, the authors establish a classification taking into account the complexity of the process and the geometry of the part to be produced. This classification makes it possible to objectively link the means to be used to obtain the parts in question.

However, the classification is currently limited to the extrusion/deposition technique, although it can be transposed or adapted to other methods used for printing cement-based materials. In order to facilitate the classification and definition of printing, the authors have developed a nomenclature that allows the definition of a type of printing.

## 1.4.2. Classification parameters

### 1.4.2.1. Scale of the printed object

The first parameter of the classification is the scale of the object to be printed. As we have seen above, the size of the object to be printed limits the range of possibilities in terms of suitable robotic systems. The authors of the classification define four different sizes which serve to inform the choice of the type of robots to be used:

– $x_0^0$: an object less than 1 meter in size;

– $x_0^1$: an object the size of a standard building element: 1–4 meters, such as a beam or slab;

– $x_0^2$: an object the size of a small building, such as a single-family home: 5–10 meters;

– $x_0^3$: an object the size of an entire building.

### 1.4.2.2. Scale of the printed material's cross-section

The second parameter for the classification is that of the scale of the cross-section of the material deposited during extrusion. This scale depends on the rheology of the material (with the height of the layer to be deposited being related to the shearing stress of the cement material; see Chapter 2), but directly influences the resolution of the printed structure, the maximum size of the cement mix's aggregates and the printing speed. The authors define four different scales:

– $x_e^0$: layer thickness less than 8 mm;

– $x_e^1$: layer thickness between 8 mm and 5 cm;

– $x_e^2$: layer thickness between 5 cm and 30 cm;

– $x_e^3$: layer thickness greater than 30 cm.

We may expect a relationship between the first two classification parameters for reasons of detail and speed quality. An optimal ratio between the size of the object to be printed and the thickness of an elementary layer must be found.

### 1.4.2.3. *Printing environment*

The authors identify three possibilities for the environment directly surrounding the printing:

– $e^0$: direct printing on-site;

– $e^1$: printing in a small mobile unit on-site, enabling control of the humidity, the temperature and the shelter from the wind;

– $e^2$: printing in a pre-fabrication plant.

The environment will directly impact the robustness of the process to be implemented. It is easier to implement the process in a protected atmosphere (as in the second or third solution) than on-site, where the changes in the weather and actions taken to correct it have to be taken into account (with regard to the protection of the printed material and the precision of the positioning of the robot). Conversely, direct on-site printing does not require handling and transport, which is an additional action to be carried out with the use of protected environments.

### 1.4.2.4. *Conditions for the assembly of the elements*

For printing in a mobile factory or in a prefabrication factory, in which it is necessary to include stages for the assembly of the various components, the authors have defined four conditions for assembling the building components that can take place during the partial or complete printing of a building:

– $a^0$: no assembly;

– $a^1$: assembly of several elements to form a larger element;

– $a^2$: handling of elements to place them in their final position;

– $a^3$: assembly of non-printed external components after printing.

It should be noted that in the case of building construction, several assembly situations may be necessary. For example, the placement of printed and non-printed elements (conditions $a^2$ and $a^4$) may take place within the same project.

### 1.4.2.5. Use of supports

As mentioned earlier, the direct manufacturing of large cantilevers is difficult, if not impossible, with the use of concrete printing by extrusion/deposition without any additional techniques. In this case, the authors of the classification have provided for the use of supports. Four categories of supports are listed as follows:

– $s^0$: no support;

– $s^1$: support printed and left in place;

– $s^2$: support printed and removed after printing;

– $s^3$: external support left in place;

– $s^4$: external support removed at the end of printing.

As with assemblies, several solutions may be used simultaneously in the printing of an object.

### 1.4.2.6. Complexity of the robot

In addition to the previous five parameters related to the form of the printed object and the printing conditions, it is also necessary to define the nature of the robot(s) used for printing. It has been conceptualized to use a collaboration between multiple robots in order to be able to print complex structures:

– $r^0$: use of a robot with three axes;

– $r^1$: use of a robot with six axes;

– $r^2$: use of a robot with six axes mounted on a rail (or mobile);

– $r^3$: use of two robots with six axes working together;

– $r^4$: use of two robots with six axes working together, mounted on rails (or mobile);

– $r^5$: use of a robot with six axes mounted on a rail (fixed or mobile);

– $r^6$: use of a robot with six axes mounted on a robot with three axes;

– $r^7$: use of two robots with six axes mounted on a robot with three axes.

Obviously, the complexity of the construction to be carried out and the first five parameters of the classification will influence the complexity of the robot configuration necessary for the printing to be done successfully. The experience and the difficulties encountered will make it possible to determine the minimum complexity of the robots used for the printing of complex structures on a large scale.

In summary, we present a classification grid that will enable readers to deduce the classification of a process, as given in Table 2.1.

## 1.4.3. *Example of classification*

In their article, the authors of the classification classify the processes used for printing according to the proposed classification parameters [DUB 17].

Following this classification, the Contour Crafting process developed by Professor Khoshnevis is classified in the following way [KHO 06]: $x_0^{0-1-2}$; $x_e^{1-2}$; $e^0$; $a^1$- $a^4$; $s^3$; $r^0$.

– Scale of the works: the process may be used to produce small buildings. It is also quite accurate, and can be used on smaller scales. It is therefore classified as $x_0^0$, $x_0^1$ or $x_0^2$, depending on the size of the object to be printed.

– Thickness of a single layer: the printed layers vary in size from 1 to 10 cm. Depending on the size chosen, the process can then be considered to be $x_e^1$, $x_e^2$.

– Environment: the Contour Crafting process involves direct manufacturing on-site. Therefore, it is $e^0$.

– Assembly: when building a single-family home, the printed elements for larger structures are put in place, and horizontal supporting elements (beams), prefabricated in other locations, are also used. Thus, the assembly steps $a^1$ and $a^4$ are used in this case.

– Use of supports: in all constructions of single-family houses, the printer uses prefabricated horizontal beams and slabs for support. These supports have a structural role and are therefore part of the construction. Thus, the condition of the supports is considered to be $s^3$. On the contrary, other parts of the printing do not need supports. They are thus considered to fall within the $s^0$ category.

– Robotic complexity: Contour Crafting uses a crane system, similar to a three-axis robot. In this case, the robotic complexity is considered as $r^0$.

| Scale of the printed object | $x_0^0$: < 1 m | $x_0^1$: 1–4 m | $x_0^2$: 5–10 m | $x_0^3$: building |
|---|---|---|---|---|
| Scale of the cross-section of the printed material | $x_e^0$ < 8 mm | $x_e^1$: 8 mm–5 cm | $x_e^2$: 5 cm–30 cm | $x_e^3$: > 30 cm |
| Printing environment | $e^0$: direct, on-site | $e^1$: control of T° (temperature) and RH (relative humidity) and 0 wind | $e^2$: prefabrication plant | |
| Conditions for the assembly of the elements | $a^0$: 0 assembly | $a^1$: assembly of several elements | $a^2$: handling of elements to place them in their final position | $a^3$: assembly of non-printed external components |
| Use of supports | $s^0$: no support | $S^1$: support printed and left in place | $s^2$: support printed and removed after printing | $s^3$: external support left in place |
| | $s^4$: external support removed at the end | | | |
| Complexity of the robot | $r^0$: three axes | $r^1$: six axes | $r^2$: six axes, mounted on a rail | $r^3$: six axes, multiple robots |
| | $r^4$: six axes, in collaboration, mounted on rails | $r^5$: six axes, mounted on a rail | $r^6$: six axes mounted on a robot with three axes | $r^7$: six axes, mounted on robots with three axes |

**Table 1.1.** *Classification of printing processes based on [DUB 17]. Parameters to be taken into account*

## 1.5. References

[ALL 16] ALL3DP, "World's First 3D Printed Pedestrian Bridge Completed in Madrid". Available at: https://all3dp.com/3d-printed-pedestrian-bridge/, December 20, 2016.

[API 18] "Apis Cor | We print buildings". Available at: http://apis-cor.com/, 2018.

[ASP 18] ASPRONE D., AURICCHIO F., MENNA C. et al., "3D printing of reinforced concrete elements: Technology and design approach", *Construction and Building Materials*, vol. 165, pp. 218–231, March 2018.

[AST 12] ASTM International, A.C.F. on A.M. Technologies and A.C.F. on A.M.T.S.F. 91 on Terminology, *Standard Terminology for Additive Manufacturing Technologies*, 2012.

[BHA 16] BHATTACHARJEE N., URRIOS A., KANG S. et al., "The upcoming 3D-printing revolution in microfluidics", *Lab on a Chip*, vol. 16, no. 10, pp. 1720–1742, 2016.

[BOS 05] BOSSCHER P., WILLIAMS R.L., TUMMINO M., "A concept for rapidly-deployable cable robot search and rescue systems", in *ASME 2005 International Design Engineering Technical Conferences and Computers and Information in Engineering Conference*, pp. 589–598, 2005.

[BOS 07] BOSSCHER P., WILLIAMS R.L., BRYSON L.S. et al., "Cable-suspended robotic contour crafting system", *Automation in Construction*, vol. 17, no. 1, pp. 45–55, 2007.

[BOS 16] BOS F., WOLFS R., AHMED Z. et al., "Additive manufacturing of concrete in construction: Potentials and challenges of 3D concrete printing", *Virtual and Physical Prototyping*, vol. 11, no. 3, pp. 209–225, July 2016.

[BOS 18] BOS F.P., AHMED Z.Y., WOLFS R.J.M. et al., "3D printing concrete with reinforcement", in *High Tech Concrete: Where Technology and Engineering Meet*, pp. 2484–2493, 2018.

[BUS 07] BUSWELL R.A., SOAR R.C., GIBB A.G. et al., "Freeform construction: Mega-scale rapid manufacturing for construction", *Automation in Construction*, vol. 16, no. 2, pp. 224–231, 2007.

[CAM 17] CAMPILLO MEJIAS M., "*Prefabricación en la arquitectura: Impresión 3D en hormigón*", Universidad Politécnica de Madrid, 2017.

[CAZ 17] "First Video of the Cazza X1 Concrete Construction 3D Printer/Video", *3D Printing Media Network*, September 20, 2017.

[CES 14] CESARETTI G., DINI E., DE KESTELIER X. et al., "Building components for an outpost on the Lunar soil by means of a novel 3D printing technology", *Acta Astronautica*, vol. 93, pp. 430–450, January 2014.

[DIL 17] DILBEROGLU U.M., GHAREHPAPAGH B., YAMAN U. et al., "The role of additive manufacturing in the era of industry 4.0", *Procedia Manufacturing*, vol. 11, pp. 545–554, 2017.

[DUB 17] DUBALLET R., BAVEREL O., DIRRENBERGER J., "Classification of building systems for concrete 3D printing", *Automation in Construction*, vol. 83, pp. 247–258, November 2017.

[GAR 11] GARDINER J., "Exploring the emerging design territory of construction 3D printing-project led architectural research", 2011.

[GOS 16] GOSSELIN C., DUBALLET R., ROUX P. et al., "Large-scale 3D printing of ultra-high performance concrete – a new processing route for architects and builders", *Materials & Design*, vol. 100, pp. 102–109, June 2016.

[GRA 19] GRAMAZIO F., KOHLER M., FLATT R.J., "Adapting Smart Dynamic Casting to Thin Folded Geometries", in *First RILEM International Conference on Concrete and Digital Fabrication–Digital Concrete 2018*, p. 81, 2019.

[GRA 97] GRAU J., MOON J., UHLAND S. et al., "High green density ceramic components fabricated by the slurry-based 3DP process", in *Solid Freeform Fabrication Symposium*, pp. 371–378, 1997.

[HAC 14] HACK N., LAUER W., "Mesh–Mould: Robotically Fabricated Spatial Meshes as Reinforced Concrete Formwork", *Architectural Design*, vol. 84, no. 3, pp. 44–53, 2014.

[HAC 15] HACK N., LAUER W., GRAMAZIO F. et al., "Mesh Mould: Robotically fabricated metal meshes as concrete formwork and reinforcement", in *Proceedings of the 11th International Symposium on Ferrocement and 3rd ICTRC International Conference on Textile Reinforced Concrete, Aachen, Germany*, pp. 7–10, 2015.

[HAC 17] HACK N. et al., "Mesh Mould: an on site, robotically fabricated, functional formwork", in *Concrete Innovation Conference HPC/CIC. Tromsø Google Scholar*, 2017.

[IZA 17] IZARD J.-B. et al., "Large-scale 3D printing with cable-driven parallel robots", *Construction Robotics*, vol. 1, nos 1–4, pp. 69–76, December 2017.

[KEA 13] KEATING S., Renaissance robotics: Novel applications of multipurpose robotic arms spanning design fabrication, utility, and art, Thesis, MIT, 2013.

[KHO 04] KHOSHNEVIS B., "Automated construction by contour crafting–related robotics and information technologies", *Best ISARC 2002*, vol. 13, no. 1, pp. 5–19, January 2004.

[KHO 06] KHOSHNEVIS B., HWANG D., YAO K.-T. et al., "Mega-scale fabrication by contour crafting", *International Journal of Industrial and Systems Engineering*, vol. 1, no. 3, pp. 301–320, 2006.

[LEC 17] LECOMPTE T., PERROT A., "Non-linear modeling of yield stress increase due to SCC structural build-up at rest", *Cement and Concrete Research*, vol. 92, pp. 92–97, February 2017.

[LIA 96] LIAO H., COYLE T.W., "Photoactive suspensions for stereolithography of ceramics", *Journal of the American Ceramic Society*, vol. 65, no. 4, pp. 254–262, 1996.

[LIA 14] LIAO X. et al., "Tower-type 3D (three-dimensional) printer and printing method thereof", CN103786235A, May 14, 2014.

[LIM 12] LIM S., BUSWELL R.A., LE T.T. et al., "Developments in construction-scale additive manufacturing processes", *Automation in Construction*, vol. 21, pp. 262–268, January 2012.

[LIN 18a] LINDEMANN H. et al., "Development of a Shotcrete 3D-Printing (SC3DP) Technology for Additive Manufacturing of Reinforced Freeform Concrete Structures", in *RILEM International Conference on Concrete and Digital Fabrication*, pp. 287–298, 2018.

[LIN 18b] LINDEMANN H., KLOFT H., HACK N., *Gradual Transition Shotcrete 3D Printing*, 2018.

[LLO 15] LLORET E. et al., "Complex concrete structures: Merging existing casting techniques with digital fabrication", *Mater. Ecol.*, vol. 60, pp. 40–49, March 2015.

[LOW 18] LOWKE D., DINI E., PERROT A. et al., "Particle-bed 3D printing in concrete construction – Possibilities and challenges", *Cement and Concrete Research*, July 2018.

[MCK 17] MCKINSEY & COMPANY, "Accélérer la mutation numérique des entreprises", p. 142, 2017.

[MON 15] MONZÓN M.D., ORTEGA Z., MARTÍNEZ A. et al., "Standardization in additive manufacturing: activities carried out by international organizations and projects", *The International Journal of Advanced Manufacturing Technology*, vol. 76, nos 5–8, pp. 1111–1121, February 2015.

[NER 16a] NERELLA V.N., KRAUSE M., NÄTHER M. et al., "Studying printability of fresh concrete for formwork free concrete on-site 3D printing technology (CONPrint3D)", in *Proceeding for the 25th Conference on Rheology of Building Materials*, 2016.

[NER 16b] NERELLA V.N., KRAUSE M., NÄTHER M. et al., "CONPrint3D–3D printing technology for onsite construction", *Concrete Institute of Australia*, vol. 42, no. 3, pp. 36–39, 2016.

[NER 17] NERELLA V.N. et al., "Micro-and macroscopic investigations of the interface between layers on the interface of 3D-printed cementitious elements", *Journal of Materials in Civil Engineering*, vol. 29, no. 7, 2017.

[NEU 16] NEUDECKER S. et al., "A new robotic spray technology for generative manufacturing of complex concrete structures without formwork", *Procedia CIRP*, vol. 43, pp. 333–338, 2016.

[PER 15] PERROT A., PIERRE A., VITALONI S. et al., "Prediction of lateral form pressure exerted by concrete at low casting rates", *Materials and Structures*, vol. 48, no. 7, pp. 2315–2322, 2015.

[PER 16] PERROT A., RANGEARD D., PIERRE A., "Structural built-up of cement-based materials used for 3D-printing extrusion techniques", *Materials and Structures*, vol. 49, no. 4, pp. 1213–1220, 2016.

[PIE 18] PIERRE A., WEGER D., PERROT A. et al., "Penetration of cement pastes into sand packings during 3D printing: Analytical and experimental study", *Materials and Structures*, vol. 51, no. 1, p. 22, January 2018.

[POU 18] POULLAIN P., PAQUET E., GARNIER S. et al., "On site deployment of 3D printing for the building construction – The case of YhnovaTM", *MATEC Web of Conferences*, vol. 163, p. 01001, 2018.

[ROU 06] ROUSSEL N., "A thixotropy model for fresh fluid concretes: Theory, validation and applications", *Cement and Concrete Research*, vol. 36, no. 10, pp. 1797–1806, 2006.

[ROU 12] ROUSSEL N., OVARLEZ G., GARRAULT S. et al., "The origins of thixotropy of fresh cement pastes", *Cement and Concrete Research*, vol. 42, no. 1, pp. 148–157, 2012.

[SHA 13] SHAHAB A.R., LLORET KRISTENSEN E., FISCHER P. et al., "Smart dynamic casting or how to exploit the liquid to solid transition in cementitious materials", in *7th RILEM International Conference on Self-Compacting Concrete and 1st RILEM International Conference on Rheology and Processing of Construction Materials*, 2013.

[SHA 17] SHAKOR P., SANJAYAN J., NAZARI A. et al., "Modified 3D printed powder to cement-based material and mechanical properties of cement scaffold used in 3D printing", *Construction and Building Materials*, vol. 138, pp. 398–409, 2017.

[VAN 19] VANDENBERG A., BESSAIES BEY H., WILLE K. et al., "Enhancing Printable Concrete Thixotropy by High Shear Mixing", in *First RILEM International Conference on Concrete and Digital Fabrication–Digital Concrete 2018*, 2019.

[WAN 16] WANGLER T. et al., "Digital concrete: Opportunities and challenges", *RILEM Technical Letters, Vol 1 2016DO – 1021809 rilemtechlett201616*, October 2016.

[WEG 16] WEGER D., LOWKE D., GEHLEN C., "3D printing of concrete structures using the selective binding method–Effect of concrete technology on contour precision and compressive strength", in *Proceedings of 11th fib International PhD Symposium in Civil Engineering*, The University of Tokyo, Tokyo, pp. 403–410, 2016.

[WIE 17] WIĘCKOWSKI A., "'JA-WA'-A wall construction system using unilateral material application with a mobile robot", *Automation in Construction*, vol. 83, pp. 19–28, 2017.

[XIA 16] XIA M., SANJAYAN J., "Method of formulating geopolymer for 3D printing for construction applications", *Materials & Design*, vol. 110, pp. 382–390, 2016.

[ZHA 13] ZHANG J., KHOSHNEVIS B., "Optimal machine operation planning for construction by Contour Crafting", *Automation in Construction*, vol. 29, pp. 50–67, January 2013.

[ZHA 18] ZHANG X. *et al.*, "Large-scale 3D printing by a team of mobile robots", *Automation in Construction*, vol. 95, pp. 98–106, November 2018.

# 2

# 3D Printing in Concrete: Techniques for Extrusion/Casting

## 2.1. Introduction

3D printing by extrusion/deposition is currently the most widely used process in the field of digital construction [WAN 16]. This automated construction process is a complex process that can be broken down into several steps: pumping the material, extruding it and depositing it. For each step of the printing process, the fresh cement-based materials must have complete control over the properties which guarantee both the flow of the material during the initial stage of the process, and its stability after being deposited. Traditionally, in laboratories, a robotic arm is used to deposit the materials transported by a mortar pump, layer-by-layer, following a path determined by the digital model of the structure to be printed (Figure 2.1). Concrete pumps are also used as a method of transporting the materials on construction sites or in prefabrication factories. Conversely, various technical solutions are used to move the extrusion (or printing) head. This is fixed on various supports that allow the effector to be moved along the programmed paths: traveling cranes [KHO 06], articulated arms on mobile machinery [LEA 12, KHO 05], fixed supports [API 18, XTR 18] or conventional building site lifting means that are repurposed from their original use [NER 16a].

---

Chapter written by Arnaud PERROT and Damien RANGEARD.

In terms of the level of complexity of the process and the consistency of the mortar or concrete to be printed, it appears that the material must have sufficient fluidity to be able to be pumped, and be firm enough to allow the stability of its form once it is deposited. These contradictory qualities require the determination of a narrow range of rheological properties in order to ensure that the printing process is carried out properly.

Another solution is to add hydration accelerators just before the deposition, through the intercalation of an in-line mixing system just upstream of the output flow. This strategy not only allows for the application of a structural build-up strategy in response to the strength requirements for the cement material, but also considers both the use of highly fluid concrete during pumping and the fast construction of firm concrete once it is deposited.

In our study, we will focus on the rheological behavior constraints induced by each stage of the extrusion/deposition printing process. First, we will give a brief overview of the process, and then we will study the transport/extrusion stage of the material. Then, we will focus our attention on the stability of a layer once it is deposited, and finally on the overall stability and control of the form once it is printed.

**Figure 2.1.** *3D printing of mortar by extrusion/deposition done in a laboratory – an articulated arm and a mortar pump. For a color version of the figures in this chapter see www.iste.co.uk/perrot/3dprinting.zip*

## 2.2. Breakdown of the process into stages

The extrusion/deposition process involves pumping a concrete mixture with an appropriate rheology and then extruding it, layer-by-layer, in order to build a structure. The process requires both a digital model and an automated system. As shown in Figure 2.2, the process can be broken down into six steps: the production of the concrete, transporting the concrete, an optional additional mixing step if a hydration accelerator is chosen, extruding the concrete, depositing a single layer, and finally the production of a structure that must be stable throughout the printing process.

There is also nothing to prevent the dry material from being transported before providing it with the water necessary for the cohesion and hydration of the cement. However, as is the case with conventional construction, dry transport is reserved for automated shotcrete construction [IBR 18].

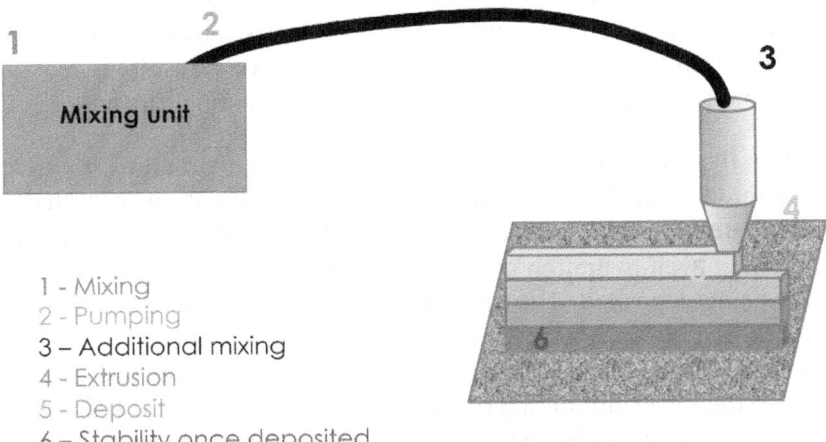

1 - Mixing
2 - Pumping
3 - Additional mixing
4 - Extrusion
5 - Deposit
6 - Stability once deposited

**Figure 2.2.** *Breakdown of the essential steps of the 3D printing of cement materials by extrusion. For a color version of the figures in this chapter see www.iste.co.uk/perrot/3dprinting.zip*

It is important to note that at each stage of the printing process, different (and sometimes opposite) physical, chemical and mechanical properties will be targeted. This gives rise to a complex formulation resulting from a knowledgeable compromise, which allows the compliance with the requirements needed for the material to undergo the various phases of printing.

This breakdown makes it possible to identify the multiple rheological properties that the cement-based material must have at each stage of the printing process. For example, it must be able to flow, while remaining homogeneous during transport and extrusion operations. In addition, the deformation of a layer during its deposition must be controlled: in this case, the elasto-plastic parameters of the mechanical behavior of the material play a decisive role in determining the scale of the project.

In addition, the quality of the interface between the layers should be treated to avoid the formation of structural weaknesses (commonly known as cold joints [WAN 16]). In the case of the last phenomenon, the cement material, which is not protected by formworks as with traditional construction, must either show resistance to drying or benefit from a compensatory water supply. As a result, it may be desirable for the mortar to have a large water retaining capacity.

Finally, the study of the stability of the printed structure at a given time should take into account the gradient of strength over the height of the structure. This gradient is induced by the different ages, resting times and even hydration degree of the material for each of the layers after being deposited. The gradients of strength are represented by the shades of gray in Figure 2.2, which highlight the importance of the kinetics of the material's structural build-up while resting. In this figure, the lower initial layers are stronger and more rigid (dark gray) than the upper layers that have just been deposited or are in the process of being deposited (light gray).

Table 2.1 summarizes the different steps of the process, as well as the effect of mechanical and physico-chemical properties on each of these steps.

|  | Mechanical/Rheological | Physical | Chemical |
|---|---|---|---|
| Mixing | Shear yield stress<br>Viscosity | Size of aggregates | – |
| Pumping | Shear yield stress<br>Viscosity | Content of aggregates<br>Size of aggregates | – |
| Additional mixing | Shear yield stress<br>Viscosity | Permeability | – |
| Extrusion | Shear yield stress<br>Viscosity | Permeability<br>Content of aggregates<br>Size of aggregates | – |
| Deposition of a layer | Shear yield stress<br>Elastic modulus | Size of aggregates | – |
| Overall stability | Shear yield stress<br>Compressive yield stress<br>Elastic modulus | Resistance to drying | Structural build-up rate<br>Setting time |

**Table 2.1.** *Interaction between mechanical, physical and chemical characteristics during the different phases of printing. The size of the aggregates should be compared with the dimensions that are used in each stage (diameter of pipes, nozzle, thickness of a layer)*

It is also of interest to define different parameters that depend on the particular printing task. Thus, the shape of the cross-section of the layers deposited (in the case of a rectangular cross-section) of height $\Delta h$ and width $w$, as well as the final height $H$ and length $L$ of the contour to be printed must be defined. Then, the advancing rate $V$ of the print head, the average rise rate $R$ and the flow rate of the material to be supplied may also be defined.

It is important to note that in the case of uninterrupted printing, it is possible to define relationships between these different parameters (e.g. between the flow rate and the advancement rate, as well as between the lift rate and the advancing rate).

*Rheology: Branch of physics that deals with the deformation and flow of matter specially the non-newtonian flow of liquids and the plastic flow of solids.*

## 2.3. Behavior during the fresh state and the printing stage

### 2.3.1. Rheology of cement-based materials

#### 2.3.1.1. Criteria for pumping material in a fresh state

When printing mortar or concrete, on the basis of the steps, the material must exhibit the behavior of a solid and hold its shape (after deposition) or the behavior of a liquid to be able to flow and be transported (during pumping and extrusion). It is therefore necessary to understand the transition between the liquid-like and the solid-like behavior of the material. This transition is a decisive characteristic of fresh concrete [ROU 08a].

In order to study in detail the steps of the printing process, it is necessary to set out the mechanical description of the behavior of the cement-based materials in a fresh state and at a very short time after being deposited.

In a fresh state, cement-based materials exhibit a complex elasto-viscoplastic behavior: they flow only when a stress greater than a critical value, known as the yield stress, is applied to them. This yield stress is observed for a critical deformation known as $\varepsilon_c$ for compression/tension or $\gamma_c$ for shear. This plastic yield stress is known as $\tau_0$ for shear flow and $\sigma_0$ for an elongation flow. For complex stress, before the cement paste or mortar is placed, a von Mises plasticity criterion may be used [MET 16] to determine whether the cement-based material will be found in the form of an elastic solid (unachieved criterion) or as a viscous fluid (achieved criterion).

This criterion can be written in the following form:

$$\sigma_{eq} = \frac{1}{\sqrt{2}}\sqrt{(\sigma_{11}-\sigma_{22})^2 + (\sigma_{33}-\sigma_{22})^2 + (\sigma_{33}-\sigma_{11})^2 + 6(\tau_{12}^2 + \tau_{13}^2 + \tau_{23}^2)}$$

$$\sigma_{eq} \leq \sigma_0 \qquad [2.1]$$

where $\sigma_{ii}$ and $\tau_{ij}$ are stress tensor components, $\sigma_{eq}$ is the von Mises equivalent stress and $\sigma_0$ is the plasticity yield stress.

Other criteria, such as the Tresca criterion, can be used. The philosophy of this criterion is identical, but the von Mises criterion is more traditionally used for concentrated suspensions such as fresh cement-based materials [PER 12, OVA 06, ADA 97].

For example, for a compression stress without being confined on axis 1, the von Mises criterion becomes

$$\sigma_{eq} = \sigma_{11}, \; \sigma_{11} \leq \sigma_0 \qquad [2.2]$$

This stress is representative of what happens when the material is deposited during printing, and is then subjected to the load from the layers placed above it.

For a simple shearing stress (axes 1 and 2), as occurs during pumping, the von Mises criterion becomes

$$\sigma_{eq} = \sqrt{3}\tau_{12}, \; \tau_{12} \leq \frac{\sigma_0}{\sqrt{3}} = \tau_0 \qquad [2.3]$$

where $\tau_0$ is the yield stress in simple shear flow. This stress is representative of what happens when the material flows through a tube, which is also known as the shear yield stress. It is this value that is traditionally used in many concrete-related manufacturing processes to qualify the consistency of the concrete and its aptness for use.

### 2.3.1.2. *Plasticity criterion not achieved: elastic behavior of a solid*

Below the yield stress (quasi-static deformation rate), it is commonly assumed that cement materials behave as linear elastic solids, with an elastic modulus $E$ in compression and $G$ in shearing. This elastic behavior is important for estimating the deformations of the structures printed under their own weight. Therefore, it is the elasto-plastic behavior that will control and scale the "static" phases of the printing, that is, after the layer has been deposited. The relationship between the stress induced by the manufacturing of the sample and those applied by the environment (wind loads, for example) and the elasto-plastic behavior should make it possible to validate the ability of a structure to be printed in a given context.

The critical elastic deformation is often used to set the limit of the elastic domain. This deformation is denoted by $\varepsilon_c$ in terms of compression/traction or $\gamma_c$ in terms of shear. The ratio of the flow yield stress, in compression and shearing, to the corresponding critical deformation allows the corresponding elastic stiffness modulus to be defined: $E = \sigma_0/\varepsilon_c$ in compression and $G = \tau_0/\gamma_c$ in shearing [ROU 18, ROU 10].

The rigidity modulus can be measured in several ways: simple compression tests, ultrasonic velocity measurement or small amplitude oscillatory "elastic" rheometry [WOL 18a, MA 18, LOO 09, PER 16]. The last two techniques, which are non-destructive, make it possible to monitor the evolution of elastic characteristics over time.

### 2.3.1.3. Plastic criterion achieved: the behavior of a viscous fluid

During the yield stress, cement-based materials flow like quasi-Newtonian viscous fluid (viscosity $\eta$) or as a power-law fluid. Therefore, in most cases (deformation rates which do not lead to loss of homogeneity), the behavior of cement materials in a wet state can be correctly approximated by a Bingham model [2.4] or by the Herschel–Bulkley model [2.5] for shearing flow (as traditionally used in the literature):

$$\tau = \tau_0 + \eta\, \dot{\gamma}, \dot{\gamma} \text{ the shear rate} \qquad [2.4]$$

$$\tau = \tau_0 + \mu\, \dot{\gamma}^n \qquad [2.5]$$

where $\tau$ is the shear stress, $\mu$ is the viscosity for the Herschel–Bulkley model and $n$ is the flow index. The value of $n$ is used to model the rheo-fluidifying behavior ($n < 1$) or rheo-thickening behavior ($n > 1$), as shown in Figure 2.3.

These models are needed to study and describe the flows involved in the pumping and extrusion of cement materials.

However, it is important to note that these models are only valid for flows established when the colloidal networks and nucleation points between cement grains are broken. In this case, the network of cement grains is considered to be unstructured. In general, rotational rheometry is used with vane-type tools to measure the parameters of the Bingham and Herschel–Bulkley models [EST 08, EST 12, FEY 12]. However, these techniques have limitations in terms of firm mixtures with shear yield stress greater than several hundred Pa [PIE 17]. In this case, a capillary geometry may be used [PER 12, ZHO 13].

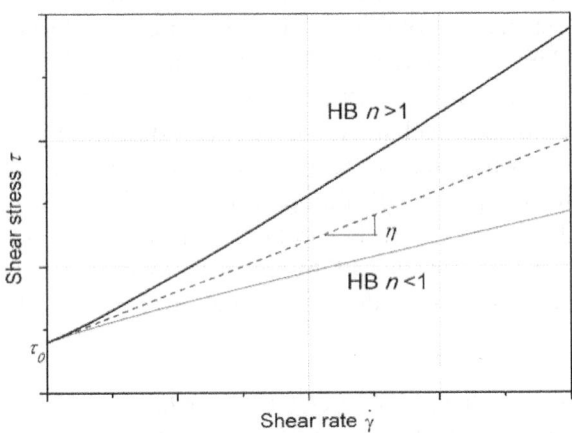

**Figure 2.3.** *Shear stress–shear rate for a Bingham fluid (linear relationship indicated by a dotted line) and two Herschel–Bulkley (HB) fluids with flow indices n less than and greater than 1*

### 2.3.1.4. *Change of rheology – physico-chemical activity over time*

Due to their nature, fresh cement materials undergo a chemical activity that changes their physical, rheological and chemical characteristics. It is this progressive structural build-up (change in characteristics) that is mobilized to allow the fresh material to build up and set in order to support the increasing load that is generated by the successive deposits of the layers forming the structure to be printed.

Thus, unlike printing using polymers, the behavior of the material cannot be considered as thermosensitive, and the stiffening of the material is not due to cooling. It is therefore necessary to rely on the kinetics of the mechanical structural build-up of the material as they relate to the activity of cement in water [WAN 16, PER 16].

To explain the stiffening of the material, it is necessary to take into account the organization of the network of cement grains in suspension in water. After an intense mixing that leads to destructuring, cement grains begin to flocculate under the influence of colloidal interactions. This flocculated lattice structure induces an increase in rigidity and strength for a period of several tens of seconds [ROU 18, ROU 12]. Then, over longer periods, the rapid formation of CSH linkages in the contact zones between the grains induces the mechanical structural build-up of the cement material left to set [ROU 12]. This phenomenon of the nucleation of the cement grains occurs during the period known as the dormant period, before setting the material (in the mechanical sense, according to Vicat [SLE 10]). These nucleations are a chemically irreversible phenomenon. However, if the power of the pumping and/or mixing system is sufficient, these links may be broken. A true loss of workability is obtained when the amount of hydrates formed is too large for the mixing system to break down all the products of the hydration reaction.

Thus, the parameters of mechanical behavior are subject to the structural build-up kinetics over time when the cement materials are left at rest after being deposited in layers. To illustrate and describe this phenomenon, the changes in the rigidity and the yield stresses are often highlighted by researchers.

The increase in the shear yield stress over time is often considered to be linear, and allows a structural build-up rate to be defined as $A_{thix}$ in $Pa.s^{-1}$ [ROU 12, ROU 06, ROU 05]. In this case, the equation can be written as:

$$\tau_0(t) = \tau_{0,t=0} + A_{thix} t \qquad [2.6]$$

where $\tau_{0,t=0}$ is the shear yield stress of the material in a destructured state and $t$ is the duration of the resting period of the material. It is possible to define a characteristic structuring time $t_{2x}$ which corresponds to the time required for doubling the material yield stress. This characteristic time has the following value:

$$t_{2x} = \tau_{0,t=0} / A_{thix} \qquad [2.7]$$

The linear modeling of shear yield stress is generally valid during the first hour of the resting period of the cement-based material [SUB 10]. Beyond that, the change accelerates and the kinetics of the structural build-up become exponential [PER 15, LEC 17]. This change in rates can be explained by the beginning of the setting process, in which the interpenetration of the CSH crystals formed can cause an increase in the solid volume fraction. Perrot et al. [PER 15] have proposed a law for the exponential increase of shear yield stress, tending towards the linear model over short periods of time. This model can be written as:

$$\tau_0(t) = A_{thix} t_c \left(e^{t/t_c} - 1\right) + \tau_{0,t=0} \qquad [2.8]$$

where $t_c$ is a characteristic time over which the behavior can be considered linear. As shown in Figure 2.4, the model of equation [2.8] can be used to describe the shear yield stress progression over longer periods. Other more sophisticated models have recently been presented in the literature [MET 16, LEC 17, WOL 18B]. Some authors have even shown that a von Mises-type plasticity criterion may ultimately not be well suited to account for the breakage of the material. Effectively, at a certain stage of the setting, the behavior of the material displays a pressure-sensitive granular type of behavior (probably related to the interconnection of the hydrates) and becomes sensitive to pressure. The mechanical behavior thus presents a progressive desymmetriszation, with a resistance that is always higher under compression than under tension. Therefore, the transition to hardened concrete behaviors begins at this time.

At the same time, critical deformation decreases slightly with the hardening of the material [MET 16, ROU 18, WOL 18B].

This increase in rigidity over time reflects an increase in the elastic modulus. These increases in rigidity and strength allow the material that is deposited to withstand the increased loads associated with the printing of the structure. Thus, it is possible to calculate and predict the optimal manufacturing speeds to guarantee the stability of the structure, and to ensure the compensation for elastic deformation, as we will see in later sections.

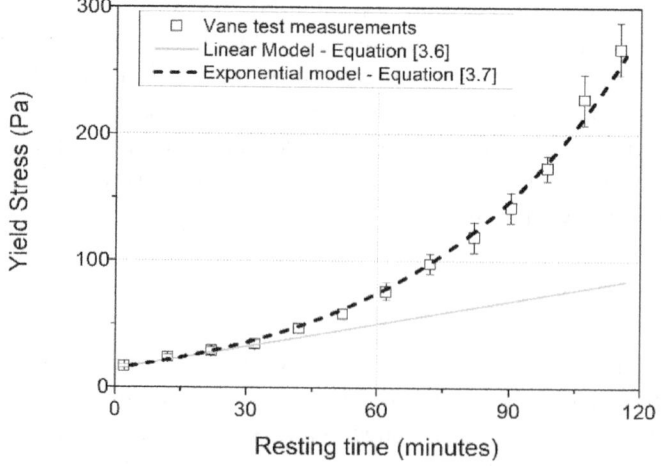

**Figure 2.4.** *Example of applications of linear and exponential models describing the shear yield stress progression of mortar left to set. The linear model will be adjusted on the basis of the measurements made within the first 30 minutes of setting*

### 2.3.2. Pumping

The pumping stage is a traditional stage in the manufacturing of concrete construction elements and other cement-based mixtures. However, this step is decisive, and may impose maximum limit characteristics in terms of shear yield stress and viscosity. These limiting characteristics may be incompatible with the stability of the printed material, which must have the minimum values of the shear yield stress to maintain its shape.

In all cases, it is recommended to adapt the power and pumping system to the formulation used. Similarly, the diameter of aggregates should be properly adapted to the operation of the pressure system and to the diameter of hose pipes. In addition, the dosage of aggregates must be limited so as not to allow friction to predominate in the flow of the cement-based material, which could lead to a blockage.

It is important to note that for pumping, delivery of the material is facilitated by the formation of a lubricating layer that consists of materials made of cement paste with a lower viscosity and yield stress (cement, water, fine particles, adjuvants) [KAP 00, LE 15, NGO 10, DES 16, FEY 13] (Figure 2.5).

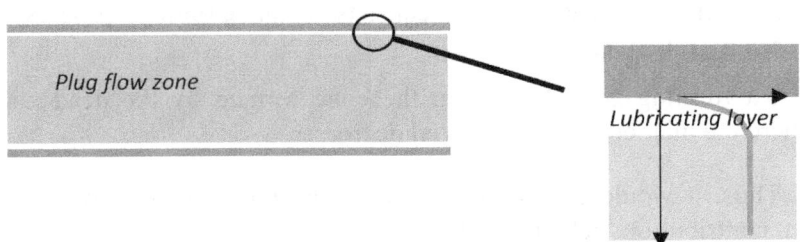

**Figure 2.5.** *Location of the lubricating layer when concrete flows through a pipe*

In this case, the study of the tribology of the material may be of interest for the description of the conditions in which this lubricating layer takes hold, and its association with the rheology of the cement-based material.

## 2.3.3. *Extrusion*

As with the pumping of cement-based materials, the flow of the material in the extruder body is often ensured by a lubricating boundary layer consisting of cement paste in which the shearing is concentrated. This type of condition generates a flow called a plug, in which the material is not sheared (or very locally, in the lubrication layers that contain a low amount of grains). Therefore, the effort generated by the extrusion is the sum of several contributions related to the breakdown of this flow into three distinct parts:

– a plug flow localized within the body of the extruder where the cement based-material appears to slide along the wall. This slipping induces a friction that must be overcome to ensure that the material flows;

– a static conical zone where the material remains blocked around the exit of the nozzle. This zone is called the dead zone. It acts as a progressive conical restriction. This zone may be limited or even non-existent if the extruder nozzle consists of an optimized progressive tapering of its shape;

– a forming zone located in the cone formed by the dead zone, where the diameter of the material decreases.

Thus, it would appear natural to divide the extrusion forces into two contributions relating to the different parts of the flow:

– the force for material forming, which has already been examined in several studies [PER 12, ZHO 13, BEN 93]. This force is denoted as $F_{pl}$;

– the force of friction $F_{fr}$, which occurs on the surface of the body of the extruder. This force is related to the wall shear stress $\tau_w$, which depends on the behavior of the lubricating layer at the interface.

Certain studies have suggested that the extrusion load is related to the shear yield stress of the material [PER 12, ZHO 13]. To show this relationship, Figure 2.6 illustrates a schematic of an axisymmetric piston extruder, and Figure 2.7 shows the corresponding extrusion force.

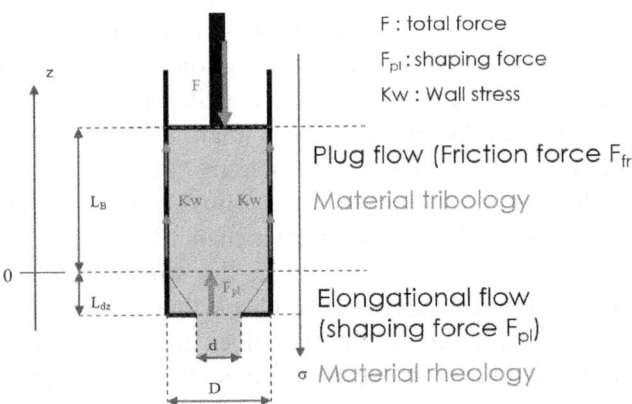

**Figure 2.6.** *Force distribution for an axisymmetric extrusion using a ram. D is the diameter of the piston, d is the diameter of the nozzle, $L_{dz}$ is the length of the conical zone and $L_B$ is the length of the plug flow zone*

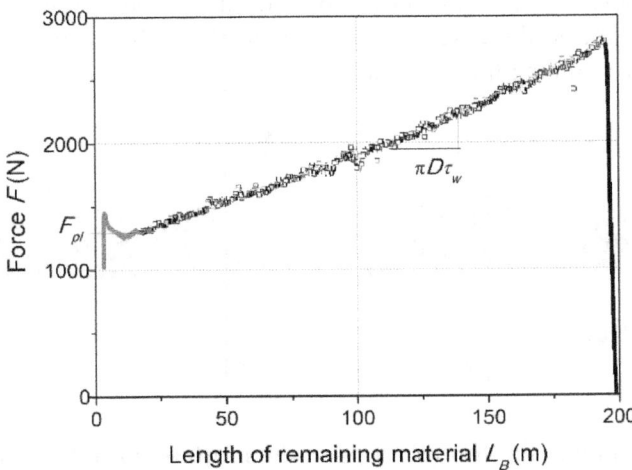

**Figure 2.7.** *Typical extrusion force progression as a function of the length of the material in the plug flow zone. After an elastic compression (black curve), the material flows, and the amount contained in the extruder decreases (decrease in linear force indicated by white dots). When the flow area plug empties, the force is equal to the plastic shaping force ($F_{pl}$)*

In all cases, for proper extrusion, it is necessary to ensure a flow without friction, and thus allow the formation of the lubricating layer. Therefore, it is not necessary to use a material with too large of a percentage of granular fraction (less than 80% of the volumetric percentage of random packing), thereby avoiding the pressure-dependent friction flow to take hold. Thus, even formulations with large yield stress (such as those greater than 1000 Pa) have been able to be printed. For the extrusion stage, the stresses are of the same order. It is nevertheless advisable to check that the pressure drop due to restriction of the section of the flow does not lead to the drainage of the interstitial fluid within the material in the pipes and the nozzle. Effectively, during the extrusion of cement-based material, a competition occurs between the kinetics of the drainage and the kinetics of the extrusion itself [PER 09, PER 14, KHE 13, TOU 05]. If the drainage occurs, the material can begin to scrape, and the surface of the layers will show more and more defects as the pressure of the pump rises. Figure 2.8 shows this drainage phenomenom during extrusion and in the case of an undrained flow (this result is obtained by simply reducing the flow rate).

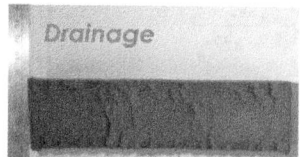 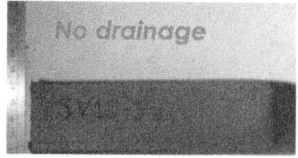

**Figure 2.8.** *View of the surface of cement materials in the case of a drained or undrained extrusion*

For example, the formulations of the mortars for 3D printing often contain a viscous agent that allows these drainage problems to be controlled or minimized.

### 2.3.4. *Stability of an elemental layer during deposition*

The deformation of a base layer after it is extruded depends on the printing process used. Following this principle, at Loughborough University [LE 12, LIM 12], small filaments in millimeter-sized cylindrical sections are extruded and deformed under their own weight. The deformation of these cylinders makes it possible to fill the gaps between filaments, thereby creating a compact structure. On the contrary, in Contour Crafting or ConPrint3D processes [KHO 04, NER 16b], the section of the layers is the same as that of the nozzle (Figure 2.9). In the first case, the final height of the material is influenced by its shear yield stress, which must be capable of withstanding the effects of gravity, and thus would be in the order of magnitude of the constraint $\rho g \Delta h$, where $\rho$ is the volumetric mass of the material and $\Delta h$ is the height of a base layer. In the second case, shear yield stress must be greater than the stress generated by gravity.

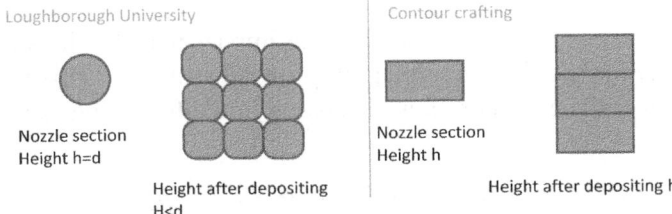

**Figure 2.9.** *Cross-section of the layers before and after deposition, and view of the different strategies for the deformation of the material after deposition*

In the case of the printing of cantilevered structures, it is recommended to verify whether the deformation remains elastic for the part of the material not supported by an underlying layer (i.e. the free part). By adopting an assumption of a cantilever beam, or using a numerical simulation by the finite element method, for example, the stability of the cantilever may be verified to make sure that the deflection of the material at the tip of the cantilever is not too excessive (Figure 2.10). This calculation of the deflection depends on the elasto-plastic behavior of the material. As the material undergoes bending and shearing, it is best to verify the elastic deformation and the lack of any plastification zones.

If the hypothesis is not verified, temporary printing support is recommended: for example, temporary soluble inks are used in the case of fused deposition modeling of polymers.

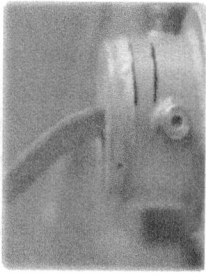

**Figure 2.10.** *View of the deformation of cantilevered mortar. The material subjected to flexural deformation shows a significant deflection as the cantilevered length increases*

It is also of interest to consider the impact that the thickness of a base layer has on the overall stability of the structure. This must take into account the considerations discussed in the following section.

### 2.3.5. *Overall stability of the printed structure in a wet state*

2.3.5.1. *Failure mechanisms for printed structures*

Initially, the ability of a cement-based material to support the weight of the structure being printed was called "buildability" or "printability" [LE 12], which could be described as "the ability of the material to be printed and to support the weight of the layers placed above it".

Since then, two causes of ruptures have been reported and analyzed in the literature. The first concerns the breakage of the base layer, which is the most heavily loaded [WAN 16, PER 16]. In this case, the failure occurs when the stress arising from the weight of the printed layers exceeds the compressive yield stress of the deposited cement material. This breakage often occurs as a result of printing structures that are not very slender [ROU 18]. The other cause for breakage that has been reported occurs due to the buckling of the printed structure [WOL 18B]. This breakage occurs as a result of printing of slender structures. The creation of cantilevered sections, or incorrect alignments, increases the risk of breakage of the printed structure due to buckling.

## 2.3.5.2. Breakage due to the collapse of the first layer

As demonstrated by Perrot et al. [PER 16], instability of the structure occurs if the strength of the material $\tau_0(t)$ becomes lower than the load acting on the first layer. This load depends on the height $h(t)$ of the structure being printed. It is therefore necessary to model both the changes over time in the mechanical strength of the cement-based material before hydration, and the mechanical load that occurs due to the construction of the structure. The theory to be developed must be able to indicate whether the layerwise structure is capable of supporting its own weight, and to predict when the structure will collapse due to the failure of the bottom layer.

During the construction of a wall or vertical column, the vertical stress acting on the first layer to be deposited increases with time as a function of the height of the structure. Even though the vertical stress gradually increases as new layers are deposited, it is possible to calculate an average rate of construction over the duration of the construction. An average vertical construction speed is determined and denoted by $R$.

Then, the vertical stress $\sigma_v$ acting on the first layer can be written as follows:

$$\sigma_V = \rho g h(t) = \rho g R t \qquad [2.9]$$

where $\rho$ is the density of the cement-based material.

The stability of the bottom layer can be assessed by comparing this vertical stress with the resistance to the compression of the material. This resistance, which according to Perrot et al. [PER 16] depends on the geometry of the printed layer, can be determined on the basis of the following relationship:

$$\sigma_c(t) = \alpha_{geom} \cdot \tau_0(t) \qquad [2.10]$$

where $\alpha_{geom}$ is a geometric factor that depends on the shape of the layers that are deposited. It is necessary to have an accurate description of the time dependent mechanical behaviour to quantify $\tau_0(t)$. As an initial approach, it is possible to use a constant structural build-up rate $A_{thix}$, as initially proposed by Roussel et al. [ROU 06, ROU 05]. In this case, the stability of the structure is controlled by the $A_{thix}/R$ ratio, which makes it possible to determine the moment at which the first layer would collapse:

$$t_f = \frac{\tau_{0,0}}{\rho g R / \alpha_{geom} - A_{thix}}$$ [2.11]

However, the linear description of the change in shear yield stress may not be sufficient to predict the effective strength of the first layer. In their work, Perrot et al. [PER 16] have essentially shown that the use of equation [2.11] could lead to the calculation of an erroneous time of failure when the shear yield stress progression accelerates (Figure 2.11).

However, it has been noted that the printing elevation speed $R$ is not necessarily an input factor into the process, but depends on other factors such as the length of the contour to be printed $L$ and the advance rate of the extruder $V$ [ROU 18]. In this case, the elevation rate depends on the geometry of the structure to be printed, which can be written as $R = V\Delta h/L$, with $\Delta h$ being the thickness of a layer.

### 2.3.5.3. Breakage by Self-buckling

Another cause of structural failure during printing that has been reported is the self-buckling of the structure. An illustration of this is given in Figure 2.12, showing the collapse via buckling of a slender structure that has been printed. This problem is theoretically connected with the theory of buckling under a structure's own weight.

In this case, in addition to the accuracy of the placing of the layers, which necessarily leads to eccentricities, which then lead to buckling instability, it is the change in the elastic modulus with resting time that controls the instability under the structure's own weight during printing [WOL 18B].

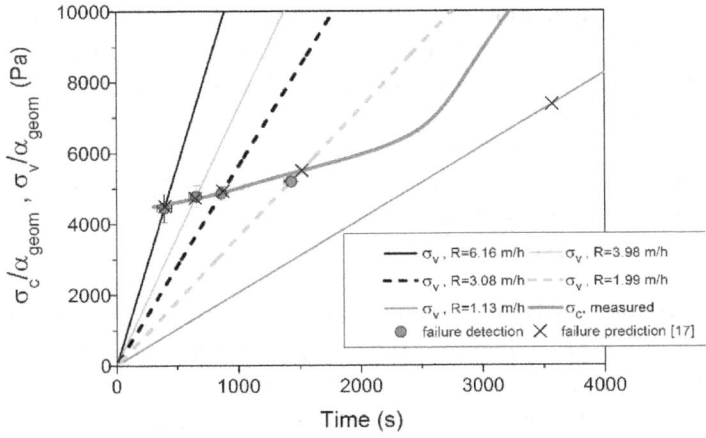

**Figure 2.11.** *Comparative changes in the resistance of the first layer compared with vertical stresses at different printing speeds. Detection and prediction of breakages with equation [2.11] (based on [PER 16])*

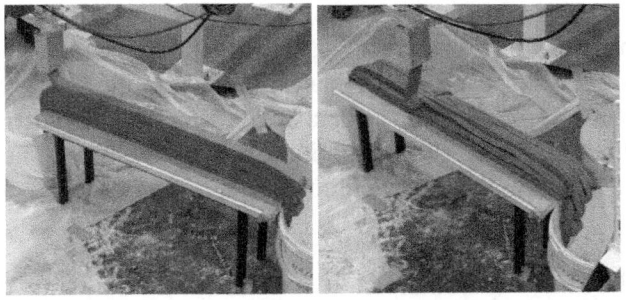

**Figure 2.12.** *Straight structure printed before and after buckling*

It should be noted that, in this case, for a strictly vertical element, the buckling limit height changes at the cube root of the printed height. Indeed, by considering a constant elastic modulus, it is possible to propose the following relationship to calculate the critical height $h_c$ resulting in the structure buckling:

$$h_c = \left(\frac{8EI}{\rho g A}\right)^{1/3} \qquad [2.12]$$

where $I$ is the quadratic moment of the printed layer and $A$ is its cross-section on a horizontal plane. Equation [2.12] demonstrates the importance of the design of the layers in minimizing the risk of buckling.

However, the elastic stiffness gradient $E$ should be considered in order to reach a more accurate prediction. It is interesting to note that buckling is the most likely type of breakage for a common wall structure (thickness 30 cm, height 2.5 m).

In addition, it is important to note that a misalignment, however small it might be, will contribute to the occurrence of the phenomenon of buckling. In the future, it would be of interest to be able to quantify dimensional tolerances that limit the risk of self-buckling.

## 2.4. Other problems occurring during concrete extrusion printing

### 2.4.1. *Elastic deformation and accuracy of the deposition*

When the beam remains stable, it will also be necessary to predict the deformation of each layer when a load is applied to it (with each layer having its own rigidity for a given time as a result of varying resting times). As a result, prediction of elastic deformation would be necessary to know the exact number of layers in order to construct a structure with a targeted height.

To illustrate this principle, it is possible to estimate the elastic deformation of the first layer deposited during printing. The values of the initial elastic modulus of the printed materials are of the level of 100 kPa [WOL 18a, PER 16, WOL 18B]. In addition, the variations in the elastic modulus are very close to those of shear yield stress, and it is possible to consider these variations in an initial approach as linear, as proposed by Roussel [ROU 06, ROU 05].

As an example, we consider the printing of a wall measuring 2.5 m high, with 50 layers of 5 cm each, using a mortar with an initial elastic modulus of 50 kPa and the speed at which the elastic modulus is increased ranging from 5 kPa/min to 20 kPa/min. The settling of the wall is calculated, as shown in Figure 2.13, with printing rates $R$ ranging from 0 to 2.5 m/h.

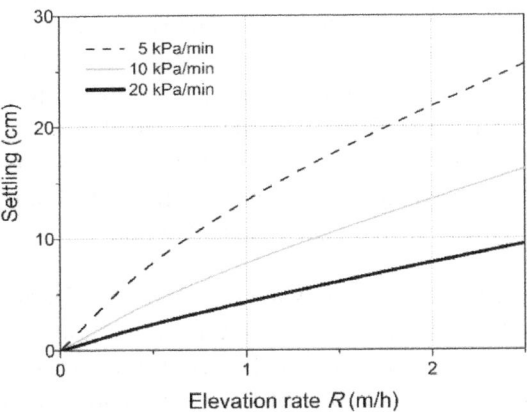

**Figure 2.13.** *Settling predicted at the top of the wall for a mortar with an increase in elastic modulus of 5 kPa/min, for the printing of a 2.5 m high wall at elevation rates ranging from 0 to 2.5 m/h*

Figure 2.13 shows that the settlement measured depends largely on the elevation rate of the elastic modulus. For printing a wall in one hour, the settlement is 25 cm for the increase in the modulus by 5 kPa/min and 9 cm for the highest speed of increasing the rigidity. In order to reduce settlement more substantially, the rate of elevation should be reduced.

Finally, the use of a setting accelerator could lead to exponential increases in the elastic modulus, which would allow more rapid hardening of the deposited layers, thereby reducing the vertical deformation of the material.

## 2.4.2. *Shrinkage and cracking during drying*

### 2.4.2.1. *Cracking due to breakage under tension*

During the printing of a curve while making a turn, the deposited material is bent along a horizontal plane. The inner part of the curve is compressed, while the outer part is stretched.

The elongation the material suffers may be greater than the critical deformation of the material, creating a crack that is detrimental not only to the aesthetic of the printed part, but also to the strength and durability of the hardened material (Figure 2.14). The risk of cracking increases as the radius of the curvature decreases, since the more abruptly the trajectory changes the direction, the greater the deformation of the external fibers becomes. Figure 2.14 shows the change in the surface of a mortar which is increasingly bent (in this case, it is given in the form of a console beam). Therefore, the greater the deformation, the greater the amount and the spacing of the cracks that appear.

It is therefore important that, when printing curved structures, the radii of the local curves should not generate a deformation greater than the critical deformation of the material. To determine this, an analysis of the strain profile of the cross-section of the mortar can be carried out in the same way as the analysis of the sections of the beam undergoing bending.

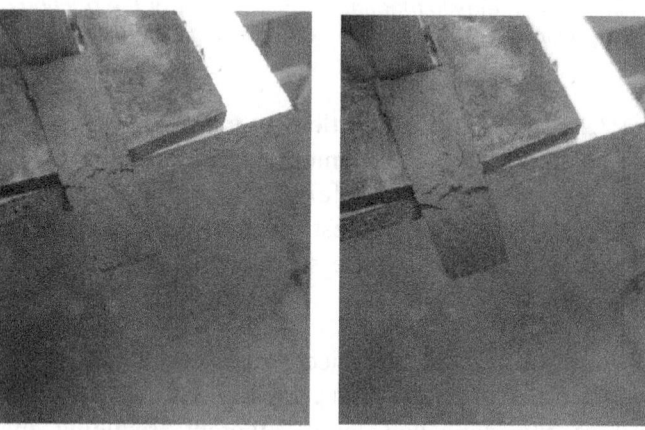

**Figure 2.14.** *Mortar cracking while undergoing flexing. Increase in the width of cracking with the increasing radii of the curvature*

### 2.4.2.2. Shrinkage and drying

Unlike concrete poured into formworks, printed structures are not protected from exterior elements while the concrete is left to set, and thus undergo drying on the surface, which can be very damaging.

The application of curing products or water-retaining agents can reduce drying and the resulting detrimental phenomena: significant shrinkage, lack of water for the cement's hydration, cracking and chipping of the surface.

It is thus essential to take into account environmental factors that have a direct effect (temperature, wind speeds, etc., as with the casting of slabs with a large surface area), in order to verify any possible drying that would be detrimental to the mechanical quality of the printed structure.

### 2.4.3. *Bonding between layers – weakness at the interface between layers*

Finally, it is necessary for the printing process to ensure proper bonding between layers and to provide proper levels of resilience to the material once it has hardened. Several solutions have been proposed in the literature, notably those made on the basis of works on multi-layered casting used for self-compacting concretes [WAN 16, ROU 08b]: forcing intermixing, trying to destructure the material in the underlying layer.

Under all circumstances, it is necessary to avoid or limit the drying of the layer that has already been deposited. For this purpose, the use of super-absorbent polymers [NER 17] and viscous agents is recommended to reduce the permeability of the material and/or its susceptibility to drying. Thus, the implementation of the wetting of the underlying layer before deposition is positive for the mechanical behavior of printed structures [SAN 18].

In all cases, the quality of the interface is important in determining the mechanical behavior of the printed material. It is important to note that a good surface quality will produce a homogeneous material that behaves as a monolithic structure, while a poor quality will produce a stratified material sensitive to delamination (Figure 2.15).

**Figure 2.15.** *Printed materials demonstrating good and bad interface qualities and the different mechanical behaviors induced. Left: test of the compression of a printed material exhibiting monolithic behavior. Right: test of the bending of a material with poor interfacing quality, inducing the phenomenon of delamination*

In all cases, it is necessary to define a maximum time limit $t_{max}$ between two successive deposits in order to ensure that the interface is of good quality, without problems of drying or mechanical connections.

### 2.4.4. Concept of time windows

It is now possible to consider the optimal period to be sought between the deposition of layers, which allows for both the stability of the structure and an absence of defects at the interface between layers.

For this purpose, we may define the printing of a structure of a height $H$ printed on layers of a thickness $\Delta h$. The number of layers can be easily deduced, and assuming a constant speed of elevation $R$, it is possible to calculate the time interval between the deposition of each layer.

In the literature, we may find limits to the value of the time interval: it must be long enough to allow the material to obtain sufficient resistance so as not to collapse, which is $t_{min}$; on the contrary, it must be short enough to avoid the formation of an interfacing defect. This second constraint provides the maximum time interval between the depositions $t_{max}$ and the minimum waiting time $t_{min}$.

If we consider the progression of the stress applied to the bottom layer and that of the strength of the material it is made from (Figure 2.16), the minimum time interval between deposits can be easily deduced from the intersection between these two curves (assuming that the structure does not break due to buckling).

To avoid the formation of cold joints, we may estimate $t_{max}$ [WAN 16, ROU 18] and compare its value with $t_{min}$. If $t_{max}$ is greater than $t_{min}$, there is no problem, and a period of time between deposits can be found within the time window defined by these two values.

Otherwise, the material is not suitable for the printing process, and the formulation should be adapted.

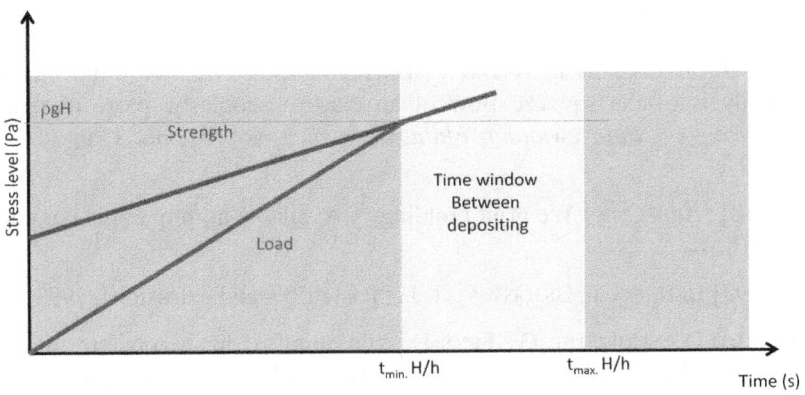

**Figure 2.16.** *View of the stress acting on the bottom layer and its resilience as a function of time. Definition of the time window between two deposits*

## 2.5. Conclusion

In this chapter, we highlighted the various aspects and phenomena involved in the printing of cement-based materials through extrusion/deposition.

Phasing the process has allowed us to list the multiple properties of cement materials in a fresh state as they are involved in the 3D printing process.

Thus, if the initial rheological behavior is involved in transporting and shaping the material, the kinetics of its progression will govern the speed of the process in order to ensure the overall stability of the structure throughout the printing.

In addition, the accuracy of the placement and the elastic modulus of the material play a role in the buckling stability and the possibility of printing cantilevered elements.

Finally, parameters of hydromechanics are also used to prevent drainage during extrusion and limit drying after deposition, and to ensure the proper quality of the interface.

## 2.6. References

[ADA 97] ADAMS M.J., AYDIN I., BRISCOE B.J. et al., "A finite element analysis of the squeeze flow of an elasto-viscoplastic paste material", *Journal of Non-Newtonian Fluid Mechanics*, vol. 71, no. 1, pp. 41–57, 1997.

[API 18] "APIS COR | We print buildings". Available at: http://apis-cor.com/, 2018.

[BEN 93] BENBOW J., BRIDGWATER J., "Paste flow and extrusion", 1993.

[DES 16] DE SCHUTTER G., FEYS D., "Pumping of fresh concrete: insights and challenges", *RILEM Technical Letters*, vol. 1, pp. 76–80, 2016.

[EST 08] ESTELLÉ P., LANOS C., PERROT A. et al., "Processing the vane shear flow data from Couette analogy", *Applied Rheology*, vol. 18, no. 3, p. 34037, 2008.

[EST 12] ESTELLÉ P., LANOS C., "High torque vane rheometer for concrete: principle and validation from rheological measurements", *Applied Rheology*, vol. 22, p. 12881, 2012.

[FEY 12] FEYS D., WALLEVIK J.E., YAHIA A. et al., "Extension of the Reiner–Riwlin equation to determine modified Bingham parameters measured in coaxial cylinders rheometers", *Materials and Structures*, vol. 46, no. 1, pp. 289–311, 2012.

[FEY 13] FEYS D., DE SCHUTTER G., VERHOEVEN R., "Parameters influencing pressure during pumping of self-compacting concrete", *Materials and Structures*, vol. 46, no. 4, pp. 533–555, 2013.

[IBR 18] IBRAHIM S. *et al.*, "Automated additive manufacturing of concrete structures without Formwork - concept for path planning", in *Tagungsband des 3. Kongresses Montage Handhabung Industrieroboter*, Springer Vieweg, Berlin, Heidelberg, pp. 83–91, 2018.

[KAP 00] KAPLAN D., "Pompage des bétons", Ecole Nationale des Ponts et Chaussées, France, 2000.

[KHE 13] KHELIFI H., PERROT A., LECOMPTE T. *et al.*, "Prediction of extrusion load and liquid phase filtration during ram extrusion of high solid volume fraction pastes", *Powder Technol.*, vol. 249, pp. 258–268, 2013.

[KHO 04] KHOSHNEVIS B., "Automated construction by contour crafting–related robotics and information technologies", *Best ISARC 2002*, vol. 13, no. 1, pp. 5–19, January 2004.

[KHO 05] KHOSHNEVIS B. *et al*, "Lunar contour crafting - a novel technique for ISRU-based habitat development", in *43rd AIAA Aerospace Sciences Meeting and Exhibit*, American Institute of Aeronautics and Astronautics, 2005.

[KHO 06] KHOSHNEVIS B., HWANG D., YAO K.-T. *et al.*, "Mega-scale fabrication by contour crafting", *International Journal of Industrial and Systems Engineering*, vol. 1, no. 3, pp. 301–320, 2006.

[LE 12] LE T.T., AUSTIN S.A., LIM S. *et al.*, "Mix design and fresh properties for high-performance printing concrete", *Materials and Structures*, vol. 45, no. 8, pp. 1221–1232, 2012.

[LE 15] LE H.D., KADRI E.H., AGGOUN S. *et al.*, "Effect of lubrication layer on velocity profile of concrete in a pumping pipe", *Materials and Structures*, vol. 48, no. 12, pp. 3991–4003, December 2015.

[LEA 12] LEACH N., CARLSON A., KHOSHNEVIS B. *et al.*, "Robotic construction by contour crafting: The case of lunar construction", *International Journal of Architectural Computing*, vol. 10, no. 3, pp. 423–438, 2012.

[LEC 17] LECOMPTE T., PERROT A., "Non-linear modeling of yield stress increase due to SCC structural build-up at rest", *Cement and Concrete Research*, vol. 92, pp. 92–97, February 2017.

[LIM 12] LIM S., BUSWELL R.A., LE T.T. et al., "Developments in construction-scale additive manufacturing processes", *Automation in Construction*, vol. 21, pp. 262–268, January 2012.

[LOO 09] LOOTENS D., JOUSSET P., MARTINIE L. et al., "Yield stress during setting of cement pastes from penetration tests", *Cement and Concrete Research*, vol. 39, no. 5, pp. 401–408, 2009.

[MA 18] MA S., QIAN Y., KAWASHIMA S., "Experimental and modeling study on the non-linear structural build-up of fresh cement pastes incorporating viscosity modifying admixtures", *Cement and Concrete Research*, vol. 108, pp. 1–9, 2018.

[MET 16] METTLER L.K., WITTEL F.K., FLATT R.J. et al., "Evolution of strength and failure of SCC during early hydration", *Cement and Concrete Research*, vol. 89, pp. 288–296, November 2016.

[NER 16a] NERELLA V.N., KRAUSE M., NÄTHER M. et al., "CONPrint3D—3D printing technology for onsite construction", *Concrete Institute of Australia*, vol. 42, no. 3, pp. 36–39, 2016.

[NER 16b] NERELLA V.N., KRAUSE M., NÄTHER M. et al., "Studying printability of fresh concrete for formwork free concrete on-site 3D printing technology (CONPrint3D)", in *Proceeding for the 25th Conference on Rheology of Building Materials*, 2016.

[NER 17] NERELLA V.N. et al., "Micro-and macroscopic investigations of the interface between layers on the interface of 3D-printed cementitious elements", *Journal of Materials in Civil Engineering*, vol. 29, no. 7, 2017.

[NGO 10] NGO T.T., KADRI E.H., BENNACER R. et al., "Use of tribometer to estimate interface friction and concrete boundary layer composition during the fluid concrete pumping", *Construction and Building Materials*, vol. 24, no. 7, pp. 1253–1261, 2010.

[OVA 06] OVARLEZ G., ROUSSEL N., "A Physical Model for the Prediction of Lateral Stress Exerted by Self-Compacting Concrete on Formwork", *Materials and Structures*, vol. 39, no. 2, pp. 269–279, 2006.

[PER 09] PERROT A., RANGEARD D., MÉLINGE Y. et al., "Extrusion criterion for firm cement-based materials", *Applied Rheology*, vol. 19, no. 5, p. 53042, 2009.

[PER 12] PERROT A., MÉLINGE Y., RANGEARD D. *et al.*, "Use of ram extruder as a combined rheo-tribometer to study the behaviour of high yield stress fluids at low strain rate", *Rheologica Acta*, vol. 51, no. 8, pp. 743–754, 2012.

[PER 14] PERROT A., RANGEARD D., MÉLINGE Y., "Prediction of the ram extrusion force of cement-based materials", *Applied Rheology*, vol. 24, no. 5, p. 53320, 2014.

[PER 15] PERROT A., PIERRE A., VITALONI S. *et al.*, "Prediction of lateral form pressure exerted by concrete at low casting rates", *Materials and Structures*, vol. 48, no. 7, pp. 2315–2322, 2015.

[PER 16] PERROT A., RANGEARD D., PIERRE A., "Structural built-up of cement-based materials used for 3D-printing extrusion techniques", *Materials and Structures*, vol. 49, no. 4, pp. 1213–1220, 2016.

[PIE 17] PIERRE A., PERROT A., HISTACE A. *et al.*, "A study on the limitations of a vane rheometer for mineral suspensions using image processing", *Rheologica Acta*, vol. 56, no. 4, pp. 351–367, 2017.

[ROU 05] ROUSSEL N., "Steady and transient flow behaviour of fresh cement pastes", *Cement and Concrete Research*, vol. 35, no. 9, pp. 1656–1664, 2005.

[ROU 06] ROUSSEL N., "A thixotropy model for fresh fluid concretes: Theory, validation and applications", *Cement and Concrete Research*, vol. 36, no. 10, pp. 1797–1806, 2006.

[ROU 08a] ROUSSEL N., COUSSOT P., OVARLEZ G., "La Pierre Liquide: des puits de potentiel au chantier", *Rhéologie*, vol. 13, pp. 41–51, 2008.

[ROU 08b] ROUSSEL N., CUSSIGH F., "Distinct-layer casting of SCC: The mechanical consequences of thixotropy", *Cement and Concrete Research*, vol. 38, no. 5, pp. 624–632, May 2008.

[ROU 10] ROUSSEL N., LEMAÎTRE A., FLATT R. J. *et al.*, "Steady state flow of cement suspensions: A micromechanical state of the art", *Cement and Concrete Research*, vol. 40, no. 1, pp. 77–84, January 2010.

[ROU 12] ROUSSEL N., OVARLEZ G., GARRAULT S. *et al.*, "The origins of thixotropy of fresh cement pastes", *Cement and Concrete Research*, vol. 42, no. 1, pp. 148–157, 2012.

[ROU 18] ROUSSEL N., "Rheological requirements for printable concretes", *Cement and Concrete Research*, May 2018.

[SAN 18] SANJAYAN J.G., NEMATOLLAHI B., XIA M. *et al.*, "Effect of surface moisture on inter-layer strength of 3D printed concrete", *Construction and Building Materials*, vol. 172, pp. 468–475, May 2018.

[SLE 10] SLEIMAN H., PERROT A., AMZIANE S., "A new look at the measurement of cementitious paste setting by Vicat test", *Cement and Concrete Research*, vol. 40, no. 5, pp. 681–686, 2010.

[SUB 10] SUBRAMANIAM K.V., WANG X., "An investigation of microstructure evolution in cement paste through setting using ultrasonic and rheological measurements", *Cement and Concrete Research*, vol. 40, no. 1, pp. 33–44, January 2010.

[TOU 05] TOUTOU Z., ROUSSEL N., LANOS C., "The squeezing test: a tool to identify firm cement-based material's rheological behaviour and evaluate their extrusion ability", *Cement and Concrete Research*, vol. 35, no. 10, pp. 1891–1899, 2005.

[WAN 16] WANGLER T. *et al.*, "Digital Concrete: Opportunities and Challenges", *RILEM Technical Letters, Vol 1 2016DO – 1021809 rilemtechlett201616*, October 2016.

[WOL 18a] WOLFS R.J.M., BOS F.P., SALET T.A.M., "Correlation between destructive compression tests and non-destructive ultrasonic measurements on early age 3D printed concrete", *Construction and Building Materials*, vol. 181, pp. 447–454, 2018.

[WOL 18b] WOLFS R.J.M., BOS F.P., SALET T.A.M., "Early age mechanical behaviour of 3D printed concrete: Numerical modelling and experimental testing", *Construction and Building Materials*, vol. 106, pp. 103–116, April 2018.

[XTR 18] "XtreeE | The large-scale 3d", available at: http://www.xtreee.eu/, 2018.

[ZHO 13] ZHOU X., LI Z., FAN M., CHEN H., "Rheology of semi-solid fresh cement pastes and mortars in orifice extrusion", *Cement and Concrete Composites*, vol. 37, pp. 304–311, 2013.

# 3

# 3D Printing by Selective Binding in a Particle Bed: Principles and Challenges

## 3.1. Introduction

Within the area of additive manufacturing of cement-based materials for the construction of elements or structures, two different techniques are used: processes based on the extrusion of fresh concrete, and the deposition of ink into layered aggregates or a powder/aggregates mixture. The ink is made up of water and additives, cement materials and/or organic materials, depending on the chosen process. These processes are commonly referred to as 3D selective binding methods for printing.

According to Claude Barlier and Alain Bernard [BAR 16], the binding of fine layers of powder using an injection of ink was patented in 1993 by Sachs [SAC 93]. Among the seven families of additive manufacturing listed in the standard ISO 17296-2, the family consisting of the projection of a binder onto a substrate, developed at MIT, is often used outside the field of civil engineering for the creation of bone implants, models for molding and various objects for art and decoration. The first part, made from concrete by selective bonding, dates back to 1995 and uses a technique in which the cement is activated by water vapor [PEG 95, PEG 97].

Chapter written by Alexandre PIERRE and Arnaud PERROT.

The principle of the technique known as additive manufacturing by selective activation is simple: a fluid material is deposited by the nozzle of a 3D printer, which then penetrates into a bed of particles to bond them together. These steps are then repeated layer-by-layer until a 3D structure that has been modeled is obtained. To do this, the dry material and the ink are moved independently at each stage, which differentiates this technique from simple concrete extrusion.

The supply system for the dry material often consists of a container placed next to the piece to be constructed. A system for leveling the particle bed can also be used to consolidate or even compact the particle bed, depending on the ink used. The final step consists of removing the solid particles that are not bonded, which can be reused to produce other parts.

The challenges of carrying out this technique more effectively in the field of civil engineering involve the choice of particular properties in terms of aggregates, mineral binder, aqueous solutions or polymers used as the binding paste/ink with regard to the need for precision, quality and fineness of the surface, the manufacturing time, and operations to be done after the production is complete. On the contrary, these techniques do not require a complete redefinition of the digital design process used in conventional additive manufacturing by the successive creation of 2D layers. The steps for creating a 3D CAD model, meshing in the STL format, and the process for the cross-sectioning and production by layers are thus suitable for production by selective bonding. Nevertheless, it is necessary to use two distinct methods for depositing the ink and powdered materials, compared with a simple extrusion.

Overcoming the scientific obstacles to the development of a reliable method of selective manufacturing in the construction sector is an important challenge in terms of technology and economics. We may recall that 30–50% of the cost of a building is established by the cost of materials and labor related to the use of casings [LLO 15]. 3D printing by selective binding allows us to envision the ability to partially eliminate the use of formworks. This technique enables us to design a lost, or a temporary, formwork without the use of wooden-based formwork and oil for removing the formworks.

In this chapter, we will present the classifications of printing by selective binding. We present an inventory of these additive manufacturing processes using selective binding, examining the different methods used as well as some of the achievements made. In this chapter, we also present the scientific challenges of obtaining a broader knowledge base with regard to the physical principles involved in the different processes used. The current theories applied for analytically modeling the printing by selective bonding are presented. The main aim of this chapter is to offer new leads for improving the control of the interaction between the materials used and the printing process, in order to make it more reliable, so as to ultimately obtain printed parts with both the mechanical behavior and the expected level of precision.

## 3.2. Classification of selective printing processes and strategies

Choosing which 3D printing technique to use involves several factors: the functionality of the final part to be made, the resolution and precision of the contours of the part, the type and durability of the material, the future load acting on the part, its size and any reinforcements it may include. Thus, the strategy to be adopted for choosing the technique to be used is determined through reflection and a design stage, which will take into account any future connections that the part will have within a construction project. The aesthetic appearance and in particular the functionality of the material produced by selective bonding are also essential in choosing the technique to be used. In the construction sector, the final product printed by selective bonding is the result of three interconnected factors: the initial material, the printer and the process it uses (speed, height of the dry material layer, etc.) and the functionality of the final part (Figure 3.1).

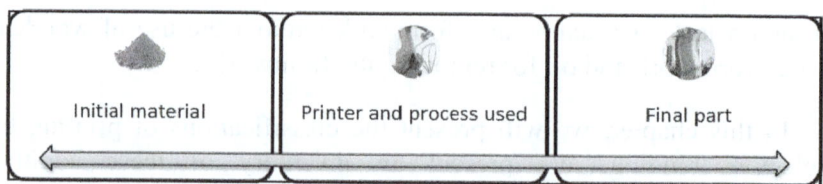

**Figure 3.1.** *Schematic breakdown of the fields considered for printing building materials by selective bonding. For a color version of the figures in this chapter see www.iste.co.uk/perrot/3dprinting.zip*

In order to realize the desired function of a part, several principles of selective manufacturing can be used in prefabrication, and in the future, on-site. According to the standard ISO 172296-2, there are seven physical principles for the addition of materials, which can be found in the book by Claude Barlier and Alain Bernard [BAR 16], in order to obtain a functional final part. Among these processes, which have been protected by patents, we will focus on those which have already been adapted to building materials.

Currently, two applications have been envisioned in the field of construction for use in production by selective binding. The first and simplest of the strategies is to directly print the final part using the method of selective activation or intrusion. The second strategy is to print the envelope of the structure or of the part of a lost formwork, and then an ad hoc reinforcement is produced by pouring concrete. It is also possible to put in place reinforcements or a reinforcement in advance. The advent of supplementary cementitious materials has created several different possibilities for producing envelopes: geopolymer, cement made up of industrial wastes, etc. In this case, the strength of the material must be just enough to support the poured concrete.

As shown in Figure 3.2, three techniques can be used to make a structure, a work or simply a part through selective production. The use of any one of these techniques can be justified by several different criteria: the type of ink, the nature of the granulate, the function of the future role played by the part in the construction work, the post-treatments, the reinforcements, etc.

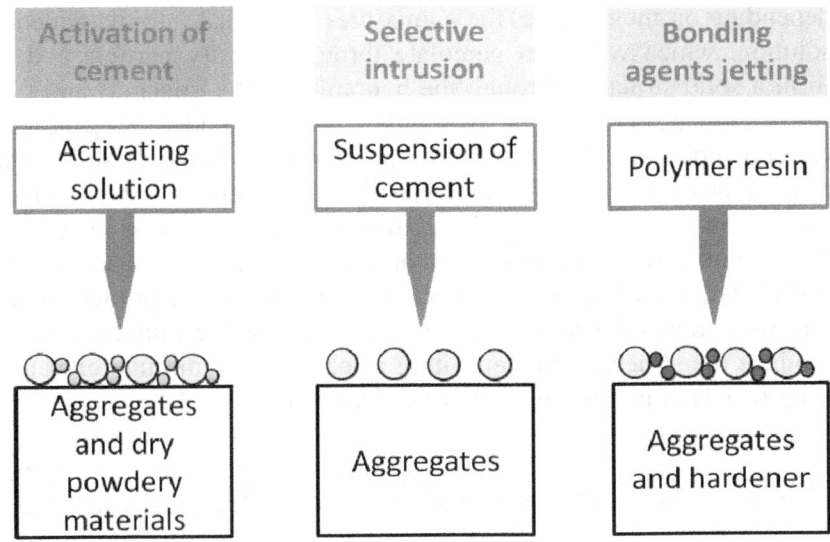

**Figure 3.2.** *Schematic breakdown of 3D printing techniques by selective manufacturing*

## 3.2.1. Selective cement activation

### 3.2.1.1. Principle

The first method, known as selective cement activation, consists of activating a powder made from cement or a powdered mineral binder mixed with aggregates (sand with or without gravel, depending on the desired printing resolution) by injecting a solution composed of water and admixtures. The powdered mineral binder is then strengthened by hydration reactions [SCR 15, MAR 18] to form a solid matrix that will encompass the aggregate particles, which are larger than the cement particles. Then, it becomes a problem of finding the right balance between the type, the size, the grain size distribution, the morphology of the powder, and the properties and the method of injection of the "activator" fluid.

Initially, the dry aggregate/sand powder and the binder are mixed together to obtain a homogeneous mixture. If the water injection is retained, then solid or liquid admixtures can be added. In general, depending on their nature, the admixtures are added into the aqueous solution, which will then percolate through the dry particle bed to form a solid structure through the hydration of the mineral binder. A sulfoaluminate cement or a highly reactive binder is generally recommended to achieve a rapid transition from liquid to solid. The printing process should also be controlled, as additives are often added during the various stages of the transporting of the ink. A setting accelerator may sometimes be incorporated once the ink reaches the end of its path, often when it reaches the nozzle of the printer. When this first composite layer is produced, two possible options exist to produce the structure: by depositing a second layer through elevation (Figure 3.3) or by placing a layer by changing the level.

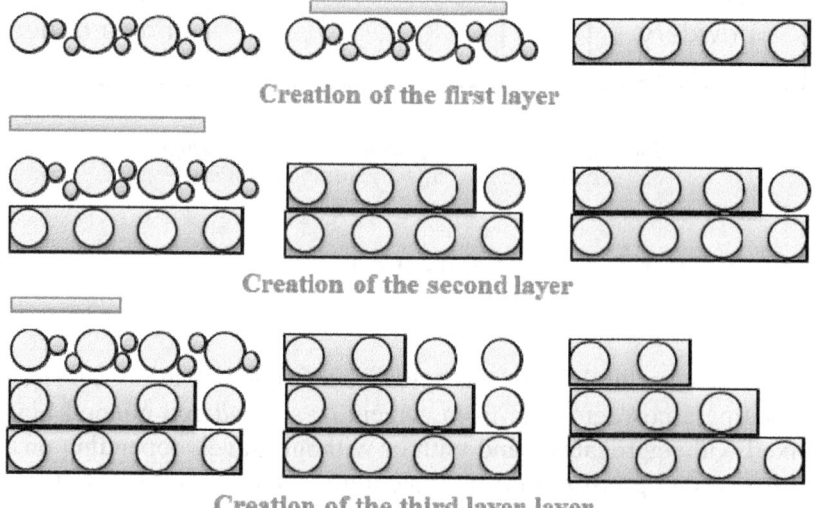

**Figure 3.3.** *Schematic of the method of selective cement activation by elevation*

### 3.2.1.2. *Applications*

Currently, the method of selective cement activation is the most widely used, notably as demonstrated in the works of Dini [CES 14] (see Figure 3.4), Lowke *et al.* [LOW 15] and Shakor *et al.* [SHA 17]. The properties of the final part are not only dependent on the formulation of the concrete, but also on the parameters of the printing process (ink viscosity, layer thickness and composition, liquid jet velocity, etc.), as reported by Lowke *et al.* [LOW 18]. The mechanical resistances in compression obtained for prismatic samples with dimensions of 40 x 40 x 160 mm$^3$ vary between 5 and 16.4 MPa, which closely correlate with the amount of water used [LOW 18]. Recent results essentially show a tendency in the opposite sense of the theory of conventional concrete formulations: the greater the amount of water used, the higher the mechanical resistance.

If the amount of water is not sufficient, it does not achieve full penetration at the height of the particle layer. In sections with low moisture, the hydration of the cement is reduced, which can result in a low level of adhesion between layers. As a result, the hydrated areas remain in the deeper parts of the particle layer, resulting in poor bonding between layers. Observations made by magnetic resonance imaging by Lowke *et al.* [LOW 15] have demonstrated the non-homogeneous distribution of water over the sample height. A high water content is found in the upper region of the layer of particles, and a low water content in the lower region.

The main obstacle to better selective control of printing through the selective activation of cement is control of the distribution of water during the printing, in order to obtain a solid interface between the layers, the complete hydration of cement particles and the homogeneity of the final part. In section 3.4, we will list the leads for new considerations on how to improve this technique.

**Figure 3.4.** *Photograph of the D-Shape 3D Printer using the cement activation method (©D-Shape)*

## 3.2.2. Selective paste intrusion

### 3.2.2.1. Principle

The selective intrusion method is simple: a fluid material is deposited by the nozzle of the 3D printer and then penetrates into the particle bed. This selective intrusion method is a process in which a suspension originating from a mineral binder locally penetrates a layer of aggregates. The suspension is often made up of a cement or mineral binding agent, water and an adjuvant or a mixture of adjuvants. In this technique of localized paste penetration, the deposited aggregate layer must be completely penetrated by the cement paste to bind with the lower layers. The cement paste must adhere to and bind the aggregate layers together to produce a homogeneous material. The homogeneity of the final part will allow a high level of resistance to be achieved during mechanical testing [LOW 18, PIE 18]. Recent results have shown that the depth of the localized penetration of the paste determines the mechanical behavior of samples manufactured by this process [LOW 18, PIE 18]. As with the selective cement activation method, it is possible to proceed by elevation (see Figure 3.5) or changing the levels to produce parts by selective paste intrusion.

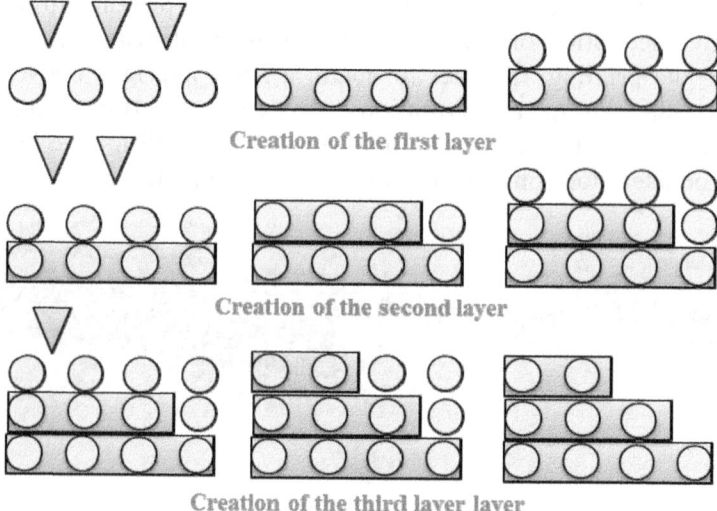

**Figure 3.5.** *Schematic breakdown of the selective paste intrusion method*

### 3.2.2.2. Applications

Currently, only a few applications are available for the use of the selective paste intrusion method. The first tests conducted by Weger *et al.* [WEG 16a, WEG 16b] showed compression strengths of 22.1 MPa on 50×50×50 mm³ samples made by selective paste intrusion with CEM I 42.5 R water to cement mass ratio W/C=0.40) in 3 mm layers. A greater level of control of the method enabled Weger *et al.* [WEG 18] to achieve compression strengths of more than 70 MPa in 7 days on cubes with sides of 100 mm.

In the case of elements produced by additive manufacturing by selective paste intrusion, the durability is unknown. Apart from the study offered by Weger *et al.* [WEG 18], only a few tests are currently available. The samples showed no carbonation after 28 days when stored with 2% $CO_2$.

**Figure 3.6.** *Photograph of tests of parts produced by selective bonding. From left to right: $d_{50}$=1, 1.6 and 2.6 mm (courtesy of Daniel Weger)*

The works of modeling and improvements made to the knowledge base on this technique can also be found in a review by Lowke *et al.* [LOW 18]. Thus, the main challenge of the selective paste intrusion technique is to control penetration of the grout used within the granular matrix. The printed structure may present discontinuities depending on the rheology of the cement paste used and the size and morphology of the particles (Figure 3.6). We will discuss recent works on this technique in section 3.4.2.

### 3.2.3. *Injection of the binder*

This technique, which is a mixture of the two previous techniques, consists of applying the binder alone to a layer made up of sand and an activator. A liquid binder, often a resin, is applied to a bed of particles in which a hardener is also included. The particle beds have a finer size than the particles used in the two methods mentioned above: selective intrusion and cement activation. When formworks are printed, the binding agent is typically a resin that reacts with a hardening component in the particle bed. Then, a construction material can be used to reinforce the structure, as shown in Figure 3.7.

**Figure 3.7.** *Sand-based mold made by the injection of a binding agent by VoxelJet, and the final part (©ETHZ)*

## 3.3. State of the art of selective printing and major achievements

The first large-scale 3D printer, the D-Shape machine, capable of producing objects with large dimensions, was built by Enrico Dini in 2005. However, we note that for printing on-site, it is necessary to use a four-axis frame, as shown in Figure 3.8. Currently, this technique is being addressed in a collaboration with ESA (European Space Agency) with the objective of forming an artificial stone corresponding to a natural dolomite stone ($CaMg(CO_3)_2$) [CES 14].

More recently, the DFAB HOUSE project in ETH Zurich [RIC 17] has developed a 3D printing technique using a powder bed to produce lost formworks, consisting of a mixture of polymer and sand.

The 3D printed element was filled with ultra-high performance fiber-reinforced concrete, and the 3D printed element surface was hardened through the infiltration of epoxy resin to seal the porous structure of the granular matrix. Figure 3.9 shows the optimization of the topology to reduce the amount of material without affecting the functionality of the beam.

**Figure 3.8.** Schematic of the D-Shape printer for on-site use, and an element of the project Underwater MOMA (©D-Shape)

**Figure 3.9.** Smart beam made by topological optimization at ETH Zurich (©Andrei Jipa, ©Hyunchul Kwon, ©Mathias Bernhard, ©Philippe Steiner)

Polymer materials, such as phenolic-based resins and particle beds with a base of PMMA particles smaller than that of sand particles, are particularly used by Voxeljet [VOX 17] for a technique using the injection of the binding agent (Figure 3.7). Geopolymers are also good candidates for the selective intrusion method, according to Xia and Sanjayan [XIA 16]. The size of the particles is therefore much finer than any sand that can be used in the production of a building element for use in a building or structure.

Currently, Portland cement is not widely used in 3D printing by selective bonding. The first use of Portland cement dates back to 1995, when Pegna [PEG 95, PEG 97] first attempted 3D printing with a particular technique using water vapor. Cement mixes were also used, such as the mixture studied by Shakor *et al.* [SHA 17]: 30% Portland cement, 65% calcium aluminate cement and 5% lithium carbonate. Portland cement with no iron oxides has also been used by the architects Rael and San Fratello [RAE 17]. It appears that the works done at the Technological University of Munich [WEG 16a, WEG 16b, WEG 18], as shown in Figure 3.10, are the precursors to printing processes by selective paste intrusion using more conventional formulations based on Portland cement CEM I, admixtures and water.

**Figure 3.10.** *Samples fabricated at the Technological University of Munich (TUM) (©Daniel Weger)*

## 3.4. Scientific challenges

### 3.4.1. *Selective cement activation and the effect of water penetration*

*3.4.1.1. Problems*

The technology making use of the projection of drops of the material as its method for activating the cement is the basis of inkjet printing technologies. The operating principle for this method is based on the deposition of drops of an aqueous solution into cement and aggregate particle beds. If a cement-based material is used, hardening is done by hydration, and penetration of the solution in the form of drops must then be controlled to ensure that a homogeneous and monolithic material is produced.

This method of selective cement activation has recently shown results that contradict the current knowledge on the effect of water on the mechanical resistance of cement-based materials. Thus, the distribution of the water is a fundamental aspect for forming a homogeneous and resistant structure. The optimized infiltration of water drops, that is, water can be found throughout the entire thickness of the layer of powder that is deposited, improves the bonding between the layers and therefore the strength of the finished product.

When the injection speed is not taken into account, the relationship between surface tension, the forces of gravity and the forces of viscosity should be compared. A recent analytical and digital modeling study [BOY 16] has presented four scenarios that may arise when a drop is deposited onto a particle bed: 1) the drop does not bond the grains due to a lack of penetration; 2) the drop remains trapped between the grains; 3) the drop descends to the bottom layer of particles; and finally 4) the drop splits and spreads around the particles (Figure 3.11).

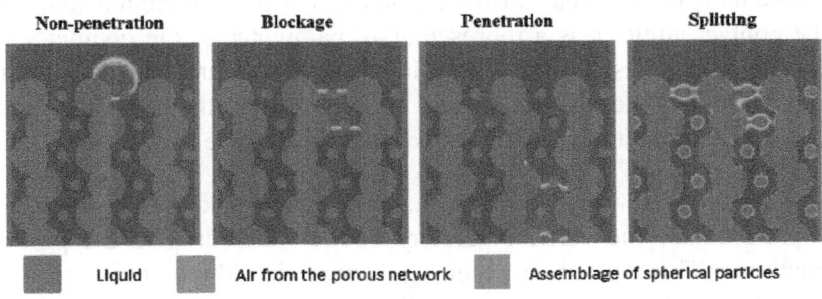

**Figure 3.11.** *Vertical cross-sections of the final liquid distribution in the simulations of the four different cases [BOY 16]*

Thus, in the future, one challenge will be to better understand the path followed by the solution as it penetrates the granular matrix. It is fair enough to raise questions about the parameters to be controlled: the injection speed, surface tension, wetting angle, the viscosity of the solution, the size of the aggregates, etc.

### 3.4.1.2. Theory

To model the one-dimensional (1D) path of a drop through the porous network of a layer of particles, the equations for its momentum can be written as follows:

$$\rho\left(\frac{\partial u_z}{\partial t} + u_z \frac{\partial u_z}{\partial z}\right) = \mu \nabla^2 u_z - \rho g - \frac{\partial p}{\partial z} \quad [3.1]$$

where $\rho$ is the density of the fluid, $g$ is the gravitational constant, $u_z$ is the vertical component of the velocity and $p$ is the pressure. If the surface tension affects the movement of the drop when it is being injected, then the pressure gradient will take into account the size of the particles, the curvature of the drop, the wetting angle, as well as the surface tension, which can be written as:

$$\frac{\partial p}{\partial z} = \frac{\beta_{bottom}\sigma\cos(\theta+\gamma_{bottom}) - \beta_{top}\sigma\cos(\theta-\gamma_{top})}{d_p^2} \quad [3.2]$$

where $d_p$ is the diameter of the particles, $\sigma$ is the surface tension, $\theta$ is the contact angle, $\beta$ is a representative parameter of the geometry of the drop and $\gamma$ is the angle of the geometry around the drop at its contact point. The top and bottom indices indicate the top and bottom contact points of the drop deposited onto the particle bed.

Equations [3.1] and [3.2] can be combined to provide an expression for the vertical velocity of the water drop in the porous medium. This speed is then governed by three properties: a contribution due to gravity, a contribution due to the surface tension of the upper part of the drop (top index) and a last contribution due to the surface tension of the lower part (bottom) of the drop:

$$u_z = \frac{d_p^2 \rho g + \beta_{bottom}\sigma\cos(\theta+\gamma_{bottom}) - \beta_{top}\sigma\cos(\theta-\gamma_{top})}{\mu} \quad [3.3]$$

This simplified model of the motion of a liquid mass of viscosity $\mu$ descending into a vertical capillary with a wavy cross-section can be used to capture the effects of the size of the particles used, the shape of the particles, as well as the effects of surface tension and wetness. This nevertheless remains a 1D model, while the real physical phenomenon is 3D.

### 3.4.1.3. *Example of the effect of the granular assembly used*

We will focus on the effect of the size of the particles on the penetration of the drops of the solution through the granular matrix, the porosity of which is assimilated into a vertical capillary. For this purpose, we will set several parameters: $\rho = 1000$ kg.m$^{-3}$, the wetting angle is set at 21°, based on Feneuil *et al.* [FEN 17], the surface tension at 70 mN.m$^{-1}$ based on Boujlel and Coussot [BOU 12], $\beta = 8$ based on Boyce *et al.* [BOY 16] and $\gamma = 45°$.

The contributions of both the effects of surface tension and gravity, scaled by their sum total, are presented in Figure 3.12. It should be emphasized that we have only considered the contributions of the bottom part, and set the effect of surface tension above zero to consider the downward flow of the drop.

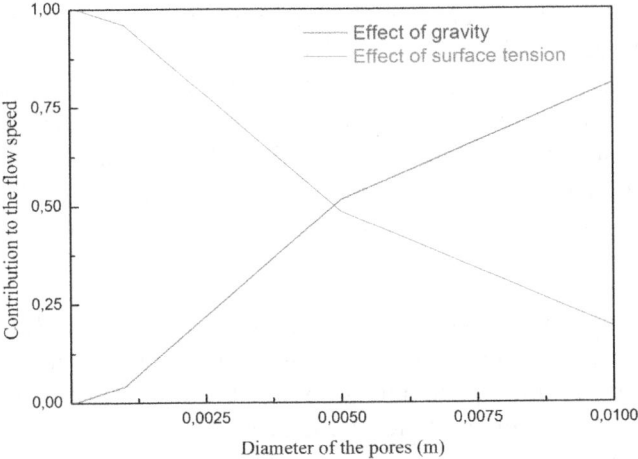

**Figure 3.12.** *Contribution of the effects of surface tension in a porous network to the downward flow rate of a drop during the cement activation process*

The results shown in Figure 3.12 indicate that the effects of surface tension are predominant up to a pore size of 5 mm, after which the flow is largely controlled by the effects of gravity. The condition of equilibrium in order for the drop to remain blocked is thus close to 5 mm, which is defined by:

$$d_p = \sqrt{\frac{\beta_{top} \sigma \cos(\theta - \gamma_{top})}{\rho g}} \qquad [3.4]$$

The application of this 1D theory (vertical velocity along a one-directional line with tortuosity), proposed by [BOY 16], demonstrates that the effects of surface tension and the angle of wetting are the parameters that will govern the path of the liquid drop within the granular formation used for the methods of cement activation. In fact, for this method, the pore sizes are well below this value of 5 mm.

It is clear from this application that the drop will be attracted towards the finer pores, and thus infiltrate horizontally due to the relatively weak influence of gravity on particles of a size less than 1 mm. Furthermore, it is unlikely that the use of surface-active agents to reduce the surface tension of the water by half [FEN 17] would be sufficient to influence the flow. In fact, the surface tension of an aqueous solution composed of surface-active agents is generally between 30 mN.m$^{-1}$ and 70 mN.m$^{-1}$ based on their composition.

The study of the geometric shape of the particles in 3D and their wetting angle should be taken into account to gain a better understanding of the phenomena involved in the activation of cement. We also note that the particle size is averaged, and that the medium is considered to be isotropic. Since the method for activating the cement is generally done using mixtures of dry powdery material consisting of micrometer-sized cement with millimeter-sized sand, proper management of the granular distribution and morphology is essential. It is thus important to model the paths a drop may take in 2D or even 3D to better understand the method of selective cement activation. Ideally, the coupling of the physical phenomena of intrusion with the physical and chemical phenomena of the hydration reactions during printing is also an interesting line of research.

Further studies should make it possible to gauge the scale of the thickness of the layers in relation to the size of the particles and the properties of the liquid injected.

### 3.4.2. *Selective intrusion and penetration by cement paste*

#### 3.4.2.1. *Problems*

In order for an element printed by selective paste intrusion to achieve a high mechanical strength, the deposited aggregate layer must be completely penetrated by the cement paste for it to bind with the lower layers. It is only when the layers are completely filled with cement paste that a monolithic and homogeneous structure with a high level of mechanical behavior properties can be made (Figure 3.13).

Therefore, the flow of cement paste through the sand layer must be predicted and controlled. As shown in Figure 3.14, the final appearance of the printed object differs depending on the grain size of the sand used for the same cement paste. The manufacturing parameters to be controlled are the rheology of the cement paste (yield stress, viscosity) and the size and morphology of the particles. We propose to study the conditions of penetration as a function of the relationship between the shearing stress of the fluid (which can be easily measured with a rheometer [OVA 11]) and the particles (which can be easily characterized), particularly in terms of their size.

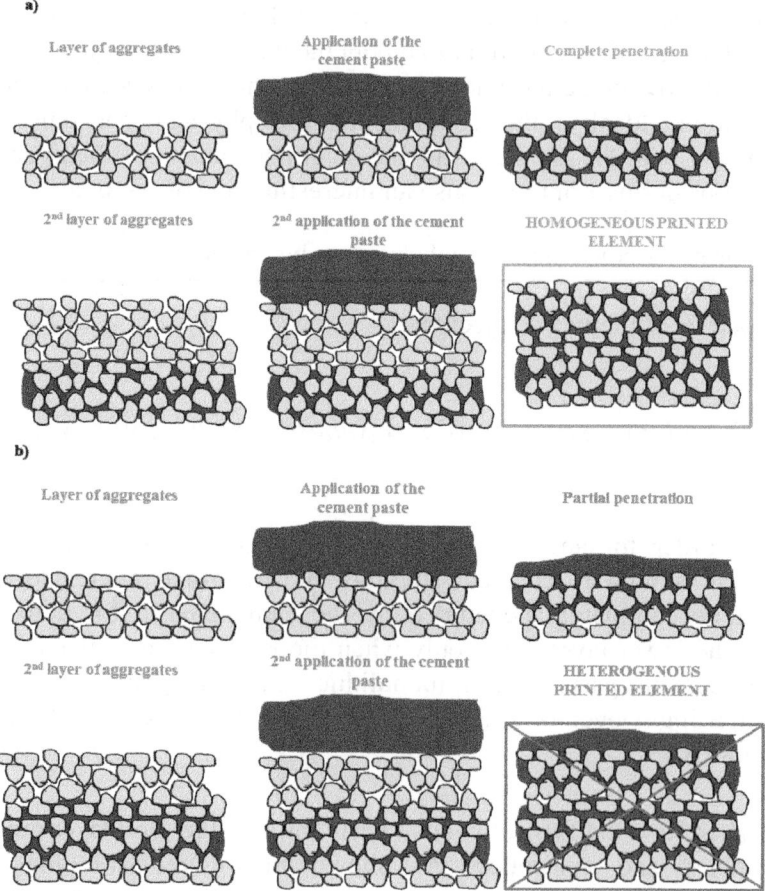

**Figure 3.13.** *Schematic of cement paste penetration: a) complete penetration; b) incomplete penetration, as demonstrated in [PIE 18]*

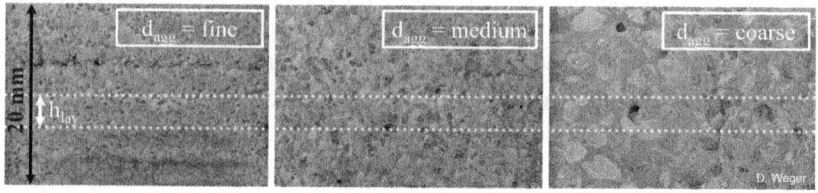

**Figure 3.14.** *Photographs of the surfaces of elements printed using the selective intrusion of a cement paste: influence of the size of the aggregate used (courtesy of Daniel Weger)*

### 3.4.2.2. Theory

To propose a model for predicting the penetration of a cement paste through a particle bed, we can consider a modified form of Darcy's law for fluids with a yield stress for their flow through a porous medium [CHE 13]:

$$D\nabla P = \alpha \tau_c + \beta k \left(\frac{V}{D}\right)^n \quad [3.5]$$

where $D$ is the average diameter of particles, $\nabla P$ is the gradient of pressure per unit length, $V$ is the velocity, $\tau_c$ is the yield stress, $k$ is the consistency and $n$ is the Herschel–Bulkley exponent. The adjustment coefficients $\alpha$ and $\beta$ of the two systems are defined independently of the rheological properties of the fluid, according to Chevalier et al. [CH 13]. The penetration of a yield stress fluid for its flow through a porous medium takes place if the following condition is met [CHE 13]:

$$D\nabla P \geq \alpha \tau_c \quad [3.6]$$

The pressure gradient is estimated according to the equations established by Green and Ampt [GRE 11, NEU 76]:

$$\nabla P(t) = \frac{1}{\Delta h(t)} \left[ \rho g \left( H(t) + \frac{v^2}{2g} \right) + \psi - \lambda(t) \right] \quad [3.7]$$

where $\rho$ is the density of the fluid, $g$ is the gravitational constant, $H$ is the height of the liquid, $v$ is the initial speed of the paste, $\psi$ is a parameter of the suction of the capillaries, $\lambda$ is a parameter of pressure loss due to friction effects and $\Delta h$ is the thickness of the porous medium to be penetrated. This parameter $\lambda$ is an essential parameter that depends on the morphology of the bed of particles, and in particular on the specific surface of the grains with which the cement paste will come into contact [PIE 18].

Using the theory developed by Pierre *et al.* [PIE 18], we can express the pressure gradient for full penetration:

$$\nabla P = \rho g - \frac{6}{d_{agg}} \frac{\Phi_S}{1-\Phi_S} \tau_c \kappa \qquad [3.8]$$

This pressure gradient must then be compared with the term of inequality, defined by [3.6].

Once penetration is completed and the velocity is zero, the maximum value of the cement paste yield stress to obtain a complete penetration can be expressed as:

$$\tau_{c,\max} = \frac{\rho g d_{agg}}{\alpha + \frac{\Phi_S}{1-\Phi_S} 6\kappa} \qquad [3.9]$$

This maximum value for yield stress depends only on the size of the particles. This criterion assumes that the penetration time is much shorter than the hardening time of the cement material to be used. Therefore, the duration of the flow must be more than 10 times shorter than the kinetics of hardening. This condition will be met if the thickness of the layer of sand to be penetrated is on the scale of tens of millimeters. Currently, the effects of thixotropy are not yet integrated into the analytical modeling of printing by selective intrusion.

It is now possible to define a final penetration height, and then a penetration rate $\Phi_{pen}$.

$$h_{pen} = \frac{\rho g d_{agg}(1-\Phi_S)h_{lay}}{\alpha\tau_c - \rho g d_{agg}\Phi_S + \frac{\Phi_S}{1-\Phi_S} 6\tau_c\kappa} \qquad [3.10]$$

$$\Phi_{pen} = \frac{\rho g d_{agg}(1-\Phi_S)}{\alpha\tau_c - \rho g d_{agg}\Phi_S + \frac{\Phi_S}{1-\Phi_S} 6\tau_c\kappa} \qquad [3.11]$$

### 3.4.2.3. *Example of the effect of the granular assembly used*

Figure 3.15 compares the penetration rates measured experimentally in a study by Pierre *et al.* [PIE 18] with those calculated via equation [3.11] given earlier.

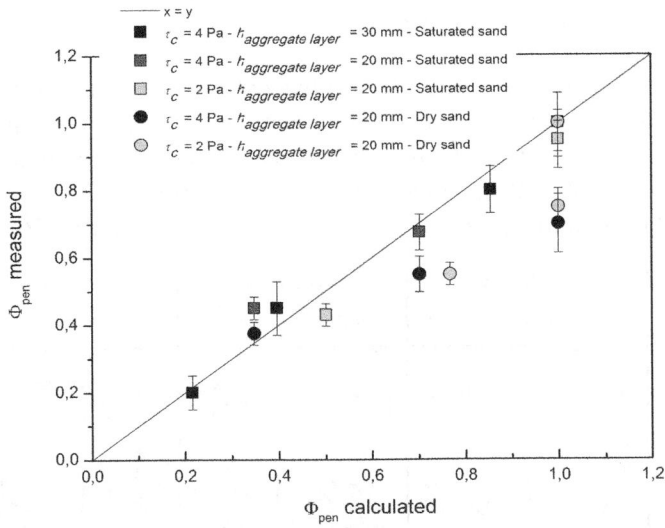

**Figure 3.15.** *Comparison of the penetration depths measured and penetration predicted by equation [3.5] of cement pastes with a yield stress of $\tau_c = 2$ Pa and $\tau_c = 4$ Pa through a layer of sand measuring 20 mm or 30 mm high, with mean diameters of 1.0, 1.6 and 2.6 mm [PIE 18]*

The results given in Figure 3.15 show that it is possible to predict the penetration of cement paste through piled sand using an analytical model, provided that the sand particles are saturated with water. When the sand is dry, the depth of the penetration measured is less than the predicted value (except for the lower yield stress and the higher particle diameter, which is the most favorable combination for penetration).

As with the method of selective cement activation, it would appear that the effects other than gravity and the rheology of the fluid may influence the height of the penetration (surface tension, wetting angle, particle morphology, water absorption by particles, etc.). The initial content of water in the sand particles is also an influential parameter. The phenomenon of the absorption of water by aggregates also plays a role in penetration, which can be studied in later works on the 3D printing method using selective intrusion of pastes.

### 3.4.3. Towards modeling in 3D

Currently, the modeling of the physical phenomena that come into play in the methods for the selective activation of the cement and the selective paste intrusion mainly focuses on 2D methods. Digital simulation tools allow relatively fast calculations to assess cement penetration within a granular network, as shown in Figure 3.16.

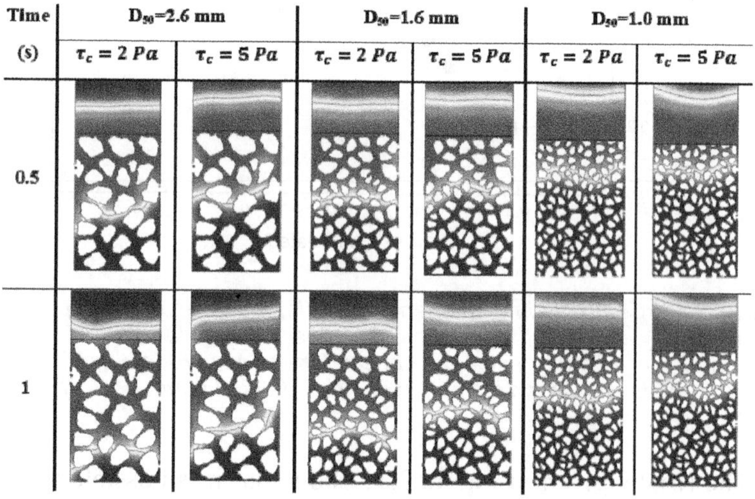

| Time (s) | $D_{50}$=2.6 mm | | $D_{50}$=1.6 mm | | $D_{50}$=1.0 mm | |
|---|---|---|---|---|---|---|
| | $\tau_c = 2\,Pa$ | $\tau_c = 5\,Pa$ | $\tau_c = 2\,Pa$ | $\tau_c = 5\,Pa$ | $\tau_c = 2\,Pa$ | $\tau_c = 5\,Pa$ |
| 0.5 | | | | | | |
| 1 | | | | | | |

**Figure 3.16.** *Tracking a cement paste's penetration into a granular mixture obtained by a numerical simulation using Comsol®*

Figure 3.16 shows the path taken by a cement paste through a granular matrix. Using this type of 2D numerical modeling, it is thus possible to estimate a penetration height and compare it to the final penetration height calculated with equation [3.10].

In reality, the path taken by the cement suspension is not solely vertical, although it is mainly governed by the effects of gravity, as demonstrated by Pierre *et al.* [PIE 18] using simple penetration tests; the device used and the results of the tests are shown in Figure 3.17.

**Figure 3.17.** *Device used to verify the nature of the downward flow due to gravity of a cement paste through a granular matrix, and the resulting material profiles (photographs taken from [PIE 18])*

We can observe that the shape of the material obtained is wider than the bottle neck used to carry out the tests. This shows that the material follows preferential flow paths when the porosity is either wider or less tortuous.

Advances in calculation clusters are expected to improve 3D digital modeling. Analytical modeling must also make advances in predicting the penetration of fluids into granular matrices, whether it is through the selective activation of cement or the selective intrusion of paste.

## 3.5. Conclusion

This chapter has presented three different strategies to encourage the use of selective printing in the construction field. We focused on two main techniques for selective binding: selective cement activation and selective paste intrusion. The achievements made using these techniques can be applied to construction elements.

It is clear that the physical phenomena involved in the techniques of printing using selective cement activation and selective paste intrusion are different. It is essentially the interaction between solids and fluids that would appear to control the penetration of solution when the cement is activated, via parameters such as the contact angle, the shape and size of the particles and the portal network, while the selective intrusion of paste is primarily controlled by gravity.

The production of homogeneous parts using selective intrusion printing is conditioned by several parameters: density, particle size, shear yield stress, etc. An initial analytical model exists, but additional consideration of the phenomena of absorption, the thixotropy of the paste and the contact angles should improve the knowledge of this technique.

It should be noted that the scientific advances that have been proposed are based on assumptions that are generally isotropic and 1D. One of the future challenges will be to propose 3D models. Effectively, the spread and the further use of digital methods for calculation will allow us to build up the knowledge base for these techniques. The positioning of sensors within the particle beds can also provide materials for digital and analytical modeling.

## 3.6. References

[BAR 16] BARLIER A., BERNARD A., *Fabrication additive du prototype rapide à L'impression 3D*, Dunod, 2016.

[BOU 12] BOUJLEL J., COUSSOT P., "Measuring yield stress: A new, practical, and precise technique derived from detailed penetrometry analysis", *Rheologica Acta*, vol. 51, pp. 867–882, 2012.

[BOY 16] BOYCE M., OZEL A., SUNDARESAN S., "Intrusion of a Liquid Droplet into a Powder under Gravity C", *Langmuir*, vol. 32, no. 34, pp. 8631–8640, 2016.

[CES 14] CESARETTI G., DINI E., DE KESTELIER X. *et al.*, "Building components for an outpost on the lunar soil by means of a novel 3D printing technology", *Acta Astronautica*, vol. 93, pp. 430–50, 2014.

[CHE 13] CHEVALIER T., CLAIN X., DUPLA J.C. *et al.*, "Darcy's law for yield stress fluid flowing through a Porous Medium", *Journal of Non-Newtonian Fluid Mechanics*, vol. 195, pp. 57–66, 2013.

[FEN 17] FENEUIL B., PITOIS O., ROUSSEL N., "Effect of surfactants on the yield stress of cement paste", *Cement and Concrete Research*, vol. 100, pp. 32–39, 2017.

[GRE 11] GREEN W.H., AMPT G., "Studies of soil physics, Part I – the flow of air and water through soils", *Journal of Agricultural Science*, vol. 4, pp. 1–24, 1911.

[LLO 15] LLORET E., SHAHAB A.R., LINUS M. *et al.*, "Complex concrete structures: Merging existing casting techniques with digital fabrication", *Computer-Aided Design*, vol. 60, pp. 40–49, March 2015.

[LOW 15] LOWKE D., WEGER D., HENKE K. *et al.*, "3D-Drucken von Betonbauteilen Durch Selektives Binden Mit Calciumsilikatbasierten Zementen – Erste Ergebnisse Zu Beton-Technologischen Und Verfahrenstechnischen Einflüssen", *Tagungsbericht '19. Internationale Baustofftagung*, P. D.-I. H.-M. Ludwig, Weimar, 2015.

[LOW 18] LOWKE D., DINI E., PERROT A. *et al.*, "Particle-Bed 3D printing in concrete construction – possibilities and challenges", *Cement and Concrete Research*, vol. 112, pp. 50–65, 2018.

[MAR 18] MARCHON D., KAWASHIMA S., BESSAIES-BEY H. *et al.*, "Hydration and rheology control of concrete for digital fabrication: Potential admixtures and cement chemistry", *Cement and Concrete Research*, vol. 112, pp. 96–110, 2018.

[NEU 76] NEUMAN S.P., "Wetting front pressure head in the infiltration model of green and ampt", *Water Resources Research*, vol. 12, no. 3, pp. 564–566, 1976.

[OVA 11] OVARLEZ G., "Caractérisation Rhéologique Des Fluides à Seuil", *Rhéologie*, vol. 20, pp. 28–43, 2011.

[PEG 95] PEGNA J., "Application of cementitious bulk materials to site processed solid freeform construction", *Solid Freeform Fabrication Symposium*, Austin, Texas, United States, 1995.

[PEG 97] PEGNA J., "Exploratory investigation of solid freeform construction", *Automation in Construction*, vol. 5, pp. 427–437, 1997.

[PIE 18] PIERRE A., WEGER D., PERROT A. et al., "Penetration of cement pastes into sand packings during 3D printing: Analytical and experimental study", *Materials and Structures*, vol. 51, no. 22, 2018.

[RIC 17] RICHARDSON V., "3D printing becomes concrete: Exploring the structural potential of concrete 3D printing", *The Structural Engineer*, pp. 10–17, 2017.

[RAE 17] RAEL R., SAN FRATELLO V., available at: www.rael-sanfratello.com, 2017.

[SAC 93] SACHS E. et al., Three-dimensional printing techniques, Bret US 5204055, 20 April 1993.

[SHA 17] SHAKOR P., SANJAYAN J., NAZARI A. et al., "Modified 3D printed powder to cement-based material and mechanical properties of cement scaffold used in 3D printing", *Construction and Building Materials*, vol. 138, pp. 398–409, 2017.

[SCR 15] SCRIVENER K., JUILLAND P., MONTEIRO P.J.M., "Advances in understanding hydration of portland cement", *Keynote Papers from 14th International Congress on the Chemistry of Cement (ICCC 2015)*, vol. 78, pp. 38–56, 2015.

[VOX 17] Voxeljet Industrial 3D Printing Systems, available at: www.voxeljet.com, 2017.

[XIA 16] XIA M., SANJAYAN J., "Method of formulating geopolymer for 3D printing for construction applications", *Materials & Design*, vol. 110, pp. 382–390, 2016.

[WEG 16a] WEGER D., LOWKE D., GEHLEN C., "3D printing of concrete structures using the selective binding method - Effect of concrete technology on contour precision and compressive strength", *11th Fib International PhD Symposium in Civil Engineering*, The University of Tokyo Edition, Tokyo, pp. 403–410, 2016.

[WEG 16b] WEGER D., LOWKE D., GEHLEN C., "3D printing of concrete structures with calcium silicate based cements using the selective binding method – effects of concrete technology on penetration depth of cement paste", *4th International Symposium on Ultra-High Performance Concrete and High Performance Construction Materials*, Kassel, Kassel University Press, 2016.

[WEG 18] WEGER D., LOWKE D., GEHLEN C., "Additive manufacturing of concrete elements using selective cement paste intrusion – Effect of layer orientation on strength and durability", *RILEM 1st International Conference on Concrete and Digital Fabrication*, Zurich, Switzerland, 2018.

# 4

# Mechanical Behavior of 3D Printed Cement Materials

## 4.1. Introduction

Given that it is a manufacturing process that is often carried out layer-by-layer, 3D printing with concrete or cement-based materials is done through the formation of a laminated material whose mechanical and physical characteristics depend on the direction in which the load falls.

If we examine the 3D printing process of extrusion/deposition that has currently been studied the most, it can be observed that the material has three planes, with orthogonal symmetry between these planes. Thus, it is an orthotropic material (Figure 4.1). Hence, its mechanical behavior differs on the basis of the three axes as defined by the direction of the deposition, the width of the layer and the height of the structure to be printed. In addition, the interfacing between layers intuitively appears to be a sensitive zone, which can have a large influence on the overall mechanical behavior of the printed material.

To fully understand the mechanical behavior of printed structures, we must take into account the level of orthotropy. This can be done by comparing its mechanical behavior in different directions of stress or

---

Chapter written by Mohammed SONEBI, Sofiane AMZIANE and Arnaud PERROT.

by comparing the characteristics of a printed material with those of the same cast material that exhibits a monolithic isotropic behavior.

In this chapter, after conducting a review of the literature on the performance of printed mortars, we will examine the possible origins of the orthotropic behavior of printed materials by considering the influence of process parameters on the origin of this nature. We will pay particular attention to the mechanical behavior of materials printed by the extrusion/deposition method. Finally, we will examine the mechanical behavior of materials printed by the other techniques.

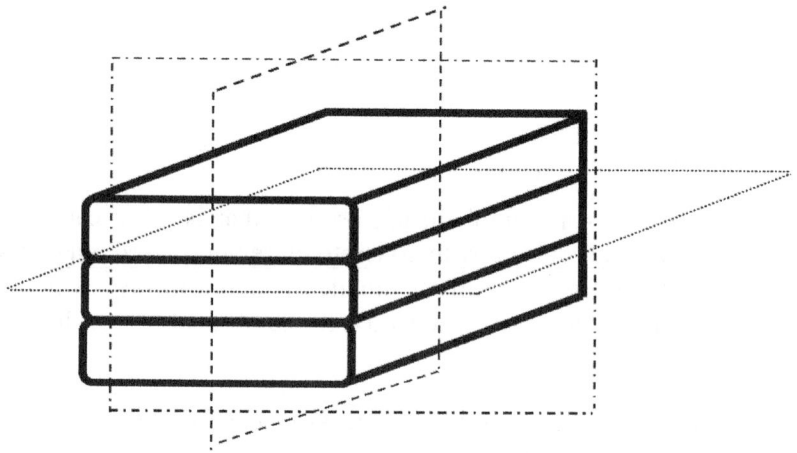

**Figure 4.1.** *Visualization of the three planes of symmetry in a sample of printed cement material*

## 4.2. Mechanical performance of the cement materials printed using the extrusion/deposition method

This chapter describes some of the results reported in the literature regarding the potential effects of the printing process through extrusion/deposition on the mechanical performance of the printed material. These effects are shown by comparing the mechanical behavior of the samples made by 3D printing and of samples made by casting (reference samples). In addition, the effect of the anisotropy of the printed concrete due to the heterogeneous elements induced by the air vacuums and the bonding strength between the layers is studied.

### 4.2.1. *Effect of extrusion on the mechanical characteristics of cement-based composites*

Extrusion is a printing process that is not very commonly used in the concrete industry. It is commonly used in the field of prefabrication, particularly high-performance fibrous technical parts (called "engineering cementitious composites" [LI 01, LI 03]) or on-site manufacturing of roadway safety barriers or sidewalk edges.

In these conventional applications, extrusion uses fresh cement-based materials with a high shear yield stress that allows the extruder to retain its shape at the end of the nozzle [PER 18]. By adding a vacuum pump to the extruder body, the extruded cement-based materials have a higher strength than the same cast material by the minimization of entrapped air bubbles [PER 06, PER 09] (Figure 4.2). Finally, it is interesting to note that an auxiliary vibration system or a lubrication system (system for the movement of water using electrical potential differences) facilitates the flow of these firm materials [PER 09, MEL 13].

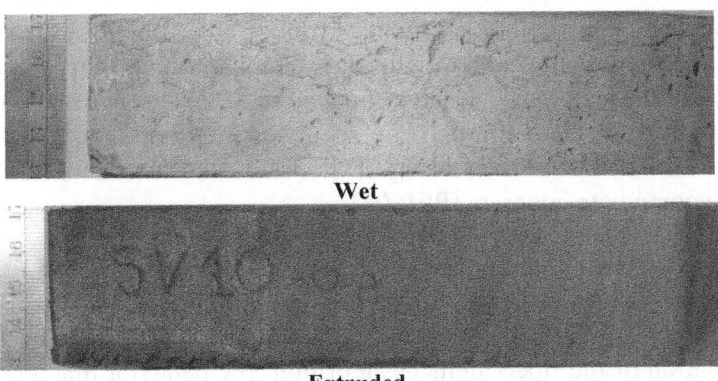

**Figure 4.2.** *View showing the reduction in the amount of entrapped air achieved through the extrusion process (water/binder mass ratio W/B = 0.23, sand/cement ratio S/C = 1, super plasticizer/cement ratio SP/C = 0.01). For a color version of the figures in this chapter see www.iste.co.uk/perrot/ 3dprinting.zip*

Several authors have studied the benefits of extrusion on the characteristics of the mechanical behavior. Thus, the mechanical performance of the same materials shaped using two different methods, the methods of casting and extrusion, were compared by Peled *et al.* [PEL 03]. The authors reported that the extrusion process improved the mechanical properties of the materials produced. These beneficial effects on extruded concrete can be partly explained by the decrease in the total porosity at both the surface and the interface of the fiber–matrix observed on the extruded samples. In this case, the matrix is stronger. The weaker characteristics of samples containing fly ash can be explained by a higher porosity obtained for these concretes. The stronger bending strength of the composite concrete was achieved by extrusion in comparison with casting [PEL 03].

In the case of the materials made by casting, with or without fly ash, the fibers failed due to tearing. With the addition of fly ash to the cast composite, the porosity increases, resulting in a decrease in the bonding between the fibers and the matrix, and thus reducing the performance of the cast composite. For the materials extruded with 2 mm PVA fibers (polyvinyl-alcohol fibers), that is, relatively short fibers, the breakage occurs through the tearing of the fibers, leading to ductile behavior, with or without the addition of fly ash. However, composites extruded with longer PVA fibers without fly ash, due to the higher adhesion that is produced (due to long fibers), exceed their critical length value and fracturing, demonstrating brittle behavior. The addition of fly ash results in an increase in the porosity and the critical length value through the reduction of fiber–matrix bonding, which contributes to the tearing of the fibers and the ductile behavior of the composite concrete [PEL 03].

In addition, the extruded fibrous composites show higher characteristics of strength, stiffness, distribution and orientation than the sample produced through casting [SHA 01]. However, the orientation of the fibers induces an orthotropic behavior that leads to a prevailing stress direction during bending (perpendicular to the direction of extrusion).

It should be noted that even extrusions done on non-fibrous cement-based materials can produce anisotropic material. In fact, the results of the measurement of the ultrasound wave velocity obtained on mortar (W/B ratio = 0.22, S/C = 1, SP/C = 0.01) show differences as a function of the directions of measurement. Thus, the velocities in the directions perpendicular to the direction of the flow were 20% lower than the velocity measured in the direction of extrusion. This can be explained by the elastic decompression of the mortar after leaving the nozzle [PER 08].

## 4.2.2. *Mechanical behavior of 3D printed cement materials*

In order to study the origin and intensity of the anisotropic (orthotropic) behavior of the printed material, we have chosen to take a two-step approach to assess this property:

1) The anisotropy induced by the multilayer printing process in comparison with the conventional casting process. In this case, the possible sources of the anisotropic behavior arise from the production of the material in layers and the quality of the interfacing.

2) The influence of adding fibers, which is intuitively likely to exacerbate the phenomenon of anisotropy due to the increased performance of each layer. In this case, the fibers are oriented longitudinally. This orientation is induced by the shear rate profile of the flow. The longitudinal orientation provides increased resistance to bending.

### 4.2.2.1. *Non-fibrous materials*

For printed concrete without fibers or reinforcements, Amziane *et al.* [AMZ 18] compared the conventional cement–sand–water–adjuvant mixture formed by a printing process using extrusion/deposition or by casting. Mortars were produced according to NF EN 1015-2 (1998) and mechanical tests were produced according to NF EN 196-1, "Methods for cement testing – Part 1: determination of resistance – Methods for cement testing" (Figure 4.3).

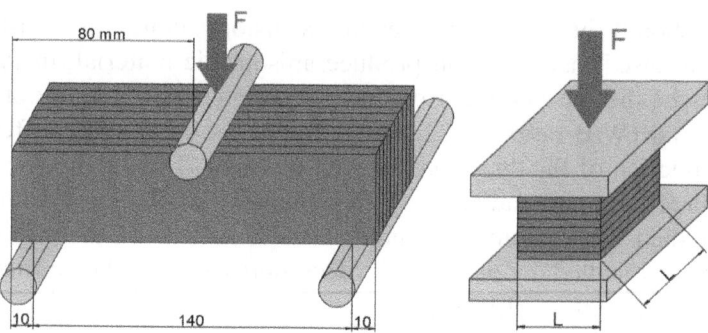

**Figure 4.3.** *Mechanical tests of flexural strength parallel to the layers and compressive strength perpendicular to the layers*

The creation of the reference mortar was developed so as to allow it to be both pumped and printed over a sufficient time period for the fabrication of the specimens, as defined in the preliminary study by Rubio *et al.* [RUB 17]. The rheology of the mortar has been optimized so that under normal circumstances, each layer retains its initial shape without collapsing under the weight of the layers placed on top of it. The reference composition of the proportions of the masses of cement, water, sand and adjuvant is 1:0.35:1.4:0.3%. The cement used is CEM II. The sand is a river type with a granulometry of 0/3 and the admixture is a modified polycarboxylate superplasticizer.

One of the first things we may learn from the study is that the printed mortars exhibit improved mechanical behavior compared with standard specimens produced through casting, in all the directions of compression and bending tests (Figure 4.4). The main reason is the positive effect of the extrusion on the density of the mortar, and the possible defects at the interfaces between the layers are largely hidden, because the printing of the layers is done quickly in this case, which does not enable the creation of cold joints.

On the contrary, we found a considerable level of heterogeneity induced by 3D printing through extrusion/deposition on the density of the mortars. The density we measured ranged from 2200 kg/m$^3$ to 2460 kg/m$^3$ from the upper layers to the lower layers of the printing. As the successive layers are printed and stacked, the lower layers begin to consolidate (Figure 4.5).

This phenomenon of consolidation has a direct impact on resistance levels, with a difference of up to 40% higher in the lowest zone with the highest consolidation, Zone N0, which is denser (Figure 4.5). However, it is important to note that the effect of consolidation is important here, because the waiting period between the two deposits is short (less than one minute) and the formulation is relatively simple (using a viscosity control agent with no specific fillers).

Tests on the parallel compression of the layers show lower strengths than those in perpendicular compression. The difference of about 15% shows that 3D printing is susceptible to the appearance of an anisotropy induced by the printing process.

In addition, Nerella *et al.* [NER 17] showed that good adhesion qualities at the interface generate quasi-isotropic behavior, close to or even better (more resistant) than that of a cast sample. This is not the case when the inter-layering is imperfect, or when the behavior of the printed material depends on the direction of the load.

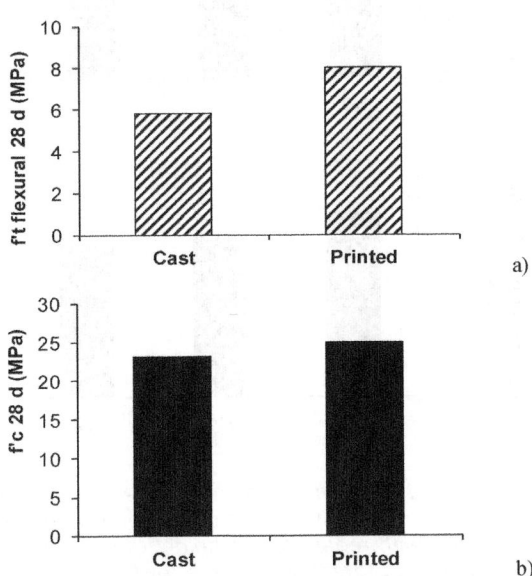

**Figure 4.4.** *The mechanical strength a) under bending and b) under compression of the reference samples (cast in accordance with EN 196-1) and printed samples using the same mortar*

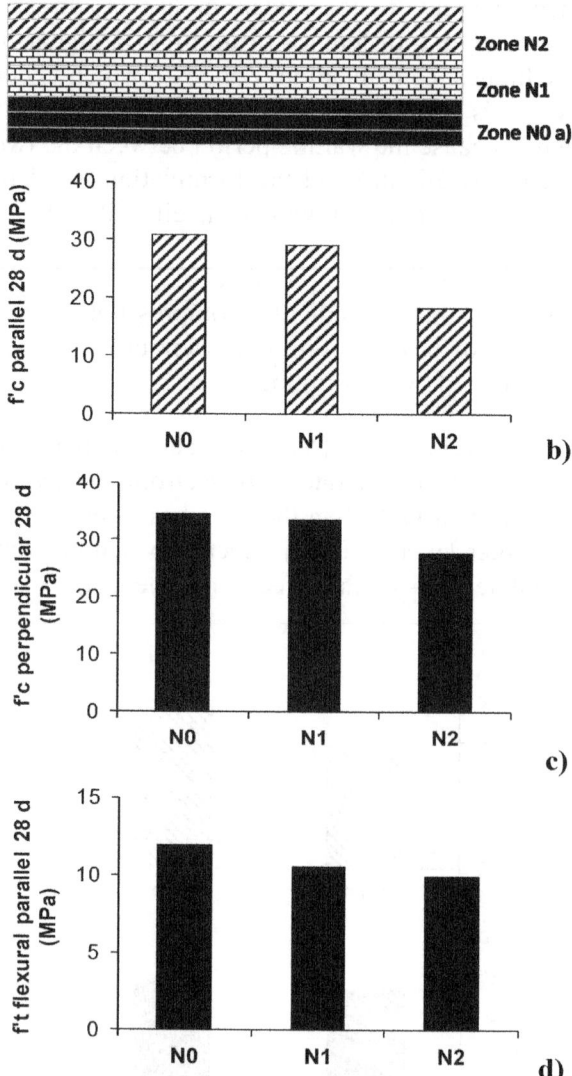

**Figure 4.5.** *Change in the strengths of a sample of nine layers: a) and b) flexural, c) compression in the layers and d) perpendicular compression, as a function of the position with regard to the height of the printed structure – the effect of the consolidation of the lower layers for the formulation is presented in Amziane et al. [AMZ 18]*

A high-performance printable mortar was developed by the same team of researchers [NER 16]. The compressive strength of samples cut by saws in directions perpendicular and parallel to the plane of the interface of the layers was 80.6 and 83.5 MPa, respectively. The results for the bending strength measurements were 5.9 and 5.8 MPa, respectively. All the results were higher than the conventionally cast samples [NER 16]. Thus, these positive results and the quasi-isotropic behavior of the material are related to the very high quality of the interface and the lack of fibers. These two parameters are the characteristics that form the basis for the anisotropy in the printed cement-based material.

These results are supported by studies by Lim *et al.* [LIM 12]. These authors found that the flexural strength represents more than 10% of the compressive strength. This study also finds that the compressive strength of extruded samples corresponds to 80–100% of the equivalent made through casting. However, the bending resistance does not differ much between the printed material and the same cast material [LIM 12]. The authors note that compressive and flexural strengths can be improved through better extrusion and by printed layers, respectively, when the anisotropy is minimized.

Similarly, Malaeb *et al.* [MAL 15] measured the compressive strength of 3D printed concrete cubes. The 3D printed concrete was made up of cement, sand and fine aggregates. The compressive strength of the samples ranged from 41.5 to 55.4 MPa, depending on the printing characteristics. This shows that the parameters of the process (printing speed, duration of the waiting period between the two successive deposits, etc.) play an important role in the mechanical behavior of the printed material. These effects will be discussed in section 4.3.2.

### 4.2.2.2. *Fibrous materials*

It has been often suggested in the literature that fibers should be used to strengthen concrete printed structures or cement-based materials [GOS 16]. For example, several authors have proposed formulations that contain fibers, to give good tensile properties to the printed cement-based materials.

Sonebi *et al.* [SON 18] proposed a new method for testing the mechanical behavior of samples of printed mortars containing natural fibers. Sets of cubes are cut into the printed structure to measure the compressive strength of layers placed above (Figure 4.6). For this purpose, five to six layers of material are printed and cut while wet in cubic molds with sides of 45 mm, and removed one minute later. In order to obtain surfaces that are perfectly parallel for the mechanical tests, the cement paste is cast to level out the surfaces to which the compression plates are applied during the tests (Figure 4.7). After 24 hours of storage at 20°C and 90% relative humidity, the samples are placed in water at (20 ±1 °C) for 7 days. The samples are tested under compression at a loading speed of 50 kN/min [SON 18].

**Figure 4.6.** *Cubic samples cut into the printed structure [SON 18]*

**Figure 4.7.** *Surfacing of cubes extracted from the printed structures [SON 18]*

Prismatic samples of dimensions 50×50×200 mm$^3$ with one or two layers are extruded and tested, as shown in Figure 4.8. The flexural test consists of a load placed on three points of the mortar samples, which are tested when they are 7 days old. The results of the multilayer samples are compared with those obtained from a monolithic sample. The samples are stored in the same way as those tested under compression [SON 18]. The loading speed was 40 kN/sec.

**Figure 4.8.** *Flexural test on a printed sample [SON 18]*

The same authors propose a test to determine the tensile strength of the printed samples in order to study the mechanical quality of the interfaces. For this purpose, a tensile strength test is carried out to determine the durability of the interface between layers up to the point of failure (Figure 4.9). The molds are filled in two steps: first, the layer is extruded into the mold. Then, the ends of the mold are filled with mortar without changing its shape. The samples are tested at a constant load speed of 1 mm/min.

**Figure 4.9.** *Test of direct tensile strength to study the mechanical quality of the interfaces [SON 18]*

Figure 4.10 shows the effects of natural fiber (NF) and silica fume (SF) on compressive strength after 7 days [SON 18]. It was found that the compressive strength of the printed layers was reduced by about 25–30% with respect to the cast cubes used for reference. The increase in the dosage of fibers did not affect compressive strength. For all the dosages of fibers and superplasticizers tested, the addition of SF resulted in a slight improvement in the compressive strength of the reference mortars, and those that were printed 7 days after their printing.

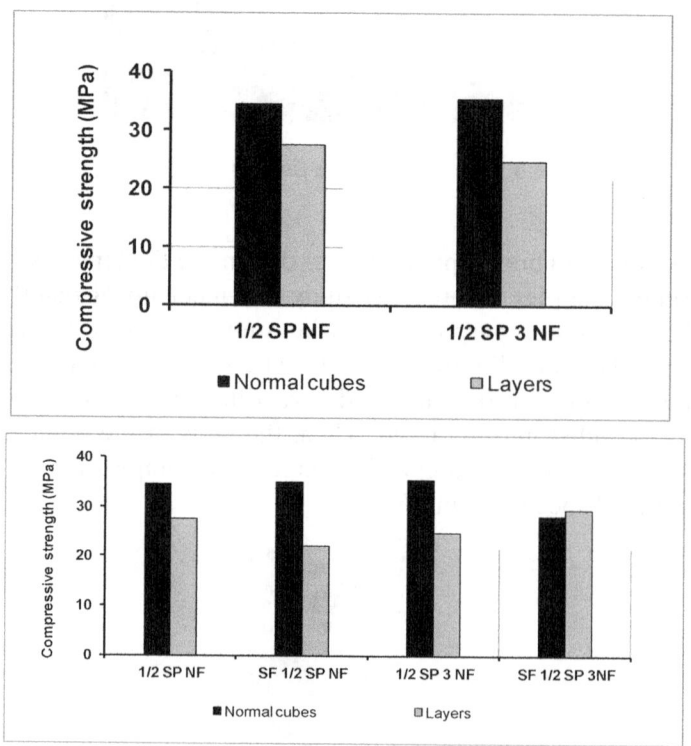

**Figure 4.10.** *Effect of the fiber and silica fume on the compressive strength of cast and printed mortar [SON 18]*

The same authors demonstrated, as shown in Figure 4.11, that the increase in the percentage of natural fibers has led to a 20% reduction in the bending resistance of a printed layer [SON 18], which is explained by the possibility that additional cracks may appear in mixtures containing more fibers. In fact, an increased fiber dosage reduced workability, and could increase air bubbles, surface defects and drainage phenomena. However, additional fibers increased the direct tensile strength by about 28%.

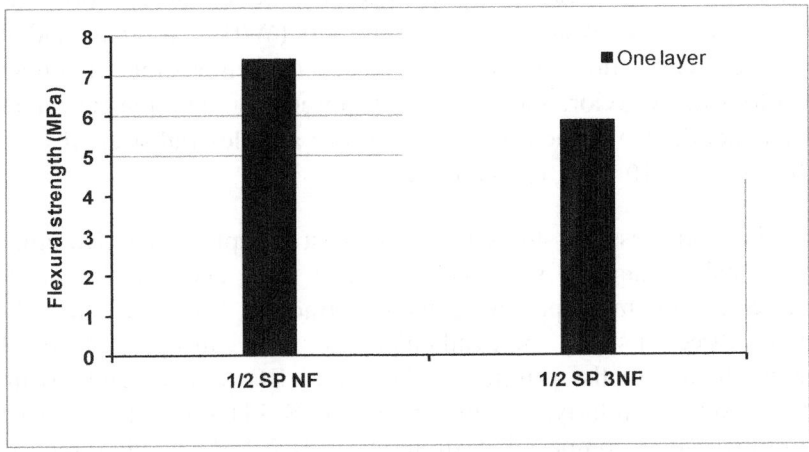

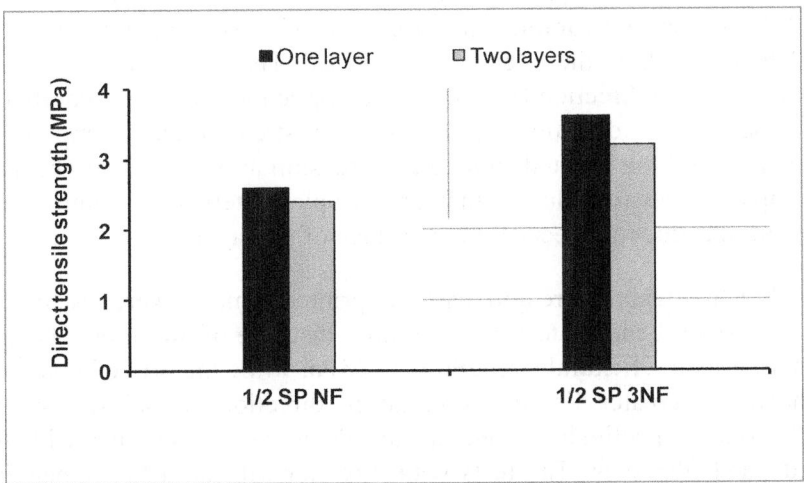

Figure 4.11. *Effect of the fiber dosage on the tensile and flexural strengths of cast and printed mortars [SON 18]*

In addition, Austin *et al.* [AUS 11] compared the effect of the anisotropic behavior of a cement-based material with a 3D printing process using extrusion with respect to samples from the same formulation made by casting. The composition of the mixture of the samples had a sand/bonding agent ratio of 60:40, consisting of 70% cement, 20% fly ash, 10% SF and 1.2 kg/m$^3$ of polypropylene microfibers. The water/binder agent ratio (W/B) was 0.26 and the water/cement ratio was 0.37. In order to demonstrate a suitable rheological behavior, the concrete required 1% superplasticizer and 0.5% retardant, and the target compressive and flexural strengths were 100 MPa and 10 MPa, respectively.

The compressive strength of the cast samples was determined using cubic samples with sides of 100 mm, and for the printed elements, 100 mm specimens were extracted. They were tested in three directions: I) perpendicular to the cutting surface; II) perpendicular to the printing surface; and III) perpendicular with a lateral side. Similarly, Austin *et al.* [AUS 11] noted that the cast samples had a compressive strength of 107 MPa in all directions, while the printed samples had a compressive strength ranging from 91 to 102 MPa in the direction of the load. The lowest resistance was obtained with direction III with a load applied to the planes parallel to the layers. In directions I and II, the strength and compressive behavior of the printed concrete were similar to those of the cast samples. They concluded that anisotropy in terms of compressive strength is due to defects at the interface of the layers.

For the flexural strength, cast and printed samples were tested with a four-point bending test device with a distance of 300 mm. The cast samples had a flexural strength of 11 MPa, while the printed materials showed a greater flexural strength in directions I and II (16 and 13 MPa, respectively) when the tensile force was aligned with the extruded filaments. The tests with direction III had a lower bending resistance due to the anisotropy of the printed cement-based materials. Essentially, the bending strength depends on the interface resistance and on the adhesion between layers, and the load was applied to the plane of the limits between the filaments. The authors highlight that printed materials have a better bending strength than conventional poured concrete [AUS 11].

The works done by the same research team enabled Le *et al.* [LE 12] to develop a high-performance 3D printed concrete containing 1.2 kg/m$^3$ of polypropylene microfibers. The values of compressive strength at 1 day, 7 days and 28 days for the 3D printing of concrete were 20, 80 and 110 MPa, respectively [LE 12]. It should be noted that these characteristics correspond to those of a very high-performance concrete.

Within this framework of research and the performances measured, Feng *et al.* [FEN 15] studied the mechanical behavior of 3D printed structures. The compressive strength of the first printed cubic samples ranged from 7.2 MPa to 16.8 MPa and was not suitable for structural elements. The authors then prepared a new, high-performance print mixture made from 30–40% cement, 40–50% crystalline silica, 10% SF and a 10% load of limestone. The results of the flexural strength of the printed samples ranged from 11.7 MPa to 16.9 MPa. To improve this, Feng *et al.* [FEN 15] produced an innovative composite structure, consisting of 3D printed elements strengthened with fiber-reinforced polymer (FRP) sheets. The maximum compressive strength of the 3D printed concrete wrapped in these FRP shells was 31.5 MPa, which was 179% higher than the non-reinforced samples [FEN 15]. Thus, the automated and selected placement of fibers or textiles can be very effective in the context of the 3D printing of cement-based materials.

Recently, Ogura *et al.* [OGU 18] showed an interest in plastic fibers to obtain materials with both high ductility and strain-hardening behavior, suitable for use in seismic zones.

In conclusion, it is possible to print concrete or mortars with very good mechanical characteristics that can be qualified as high or very high performance. Anisotropy can be found in fibers oriented by the deposition during extrusion and by the poor quality of the interface between layers.

## 4.3 Effects of the additive fabrication method on the mechanical behavior of cement-based materials

### 4.3.1. *Printed concrete = anisotropic stratified materials: possible causes*

As mentioned in the introduction, production by layering can lead to orthotropy, due to the presence of joints at the interface between possibly imperfect layers. Indeed, imperfect contacts between the layers, the failure of the material to mix with the inter-layer or drying that may occur during the period between two successive layer deposits can lead to a heterogeneity of the material across its height, which leads to the lack of isotropy of the material.

Similarly, as we have discussed above, the transporting, extrusion or deposition may result in prevailing orientations of defects and particles in the direction of the flow. This is particularly true in the case of fibrous materials where the fibers tend to orient themselves in the direction of the flow. Thus, it results from the quality of the interface between layers and the transport of the material. The mechanical behavior of the printed materials will be more or less isotropic.

### 4.3.2. *Effects of the printing process parameters on the mechanical properties*

#### 4.3.2.1. Effects of the time intervals between successive deposits

In his article on the state of the art of 3D printing, Wangler *et al.* [WAN 16] indicate that an excessively long waiting time between deposits could lead to defects in bonding at the interface areas, because of insufficient mixing between layer materials, referred to as "cold joints". The mixing of materials between layers is ensured on the stress levels related to the speed of the deposition of the material in excess of the shear yield stress of the layer already in place. This theory has been borrowed from the theory of the appearance of casting defects found between self-consolidating concrete castings [ROU 10]. Given the strength of cementitious materials that build up over time (the phenomenon of thixotropy of yield stress), it is thus necessary not to wait too long in order to ensure a proper remixing. The application of this theory works well when the deposited material behaves in a manner close to that of a self-consolidating concrete.

For materials with a higher initial shear yield stress, an illustration of the effect of the time between deposition is given by the experiment on mortar with an initial shear threshold of 2 kPa. Cast and three-layer printed samples were manufactured with deposition times between layers ranging from 1 to 10 minutes (dimensions of 30×36×160 cm$^3$). The samples were then tested under bending and compression. The results obtained (Table 4.1) show a decrease in flexural and compressive strengths over the duration between deposits. In addition, the samples printed with a time period of one minute between the deposition exhibit a monolithic behavior, whereas the specimens printed with a time period of 10 minutes between deposition exhibit cracking at the interfaces (Figure 4.12).

|  | Printed 1 minute | Printed 2 minutes | Printed 5 minutes | Printed 10 minutes | Test samples Cast |
|---|---|---|---|---|---|
| Compression (MPa) | 43.5 | 34.5 | 32.2 | 23.1 | 46.5 |
| Flexural (MPa) | 4.30 | 4.3 | 4.0 | 3.70 | 5.1 |

**Table 4.1.** *Compressive resistances of cast and printed test samples after different periods between deposits*

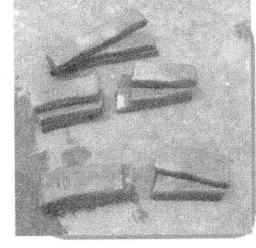

Wait time 1 minute          Wait time 10 minutes

**Figure 4.12.** *Breakage of printed test samples under bending with different waiting times between deposits. Monolithic behavior for the lowest waiting time – appearance of weaknesses at the interfaces and delamination for the longest waiting time*

These defects also occur due to changes in the drying conditions on the surface (drying). Several studies have shown this influence. We will examine this further in the next section.

### 4.3.2.2. Effects of water content – detrimental effects of drying

Sanjaya *et al.* [SAN 18] have shown that the hydric state of the surface of the material has a strong influence on the behavior of the hardened material. In this sense, the authors show that drying reduces the surface water content and negatively affects the mechanical properties of the material. They also show that the presence of water on the surface, which may result from water bleeding outward, has a positive effect on the material.

In addition, Nerella and his co-authors have suggested that the use of microsilica reduces the porosity of the deposited materials and thus allows a quasi-constant water content to be maintained, and thus maintains a good quality between the layers [NER 16, NER 17]. Similarly, the use of adjuvants, such as superabsorbent polymers or viscous agents such as cellulose ethers, could also have a beneficial effect on the quality of interfacing.

To illustrate the effect of the influence of the water content, water was applied in a mist with a spraying device just before the deposition of a new layer on the materials, as shown in Figure 4.12. The results obtained showed that by spraying the water on the layer in place allowed the reduction of compressive strength to be decreased compared to those obtained on samples not wetted. On the contrary, a monolithic behavior was observed, even for the longest waiting times tested (Table 4.2).

|  | Printed 1 minute | Printed 2 minutes | Printed 5 minutes | Printed 10 minutes | Test samples Cast |
|---|---|---|---|---|---|
| Compression with wetting (MPa) | 46 | 41.5 | 39.2 | 31.1 | 46.5 |
| Compression without wetting (MPa) | 43.5 | 34.5 | 32.2 | 23.1 | 46.5 |

**Table 4.2.** *Compressive strength of cast and printed test samples after different periods between deposits*

## 4.4. Mechanical behavior obtained with other methods of 3D printing of cement-based materials

### 4.4.1. *Production using robotic sliding castings ("Smart Dynamic Casting")*

For the Smart Dynamic Casting process developed at ETH Zurich, the behavior of the material is that of conventional self-consolidating concrete [LLO 15]. In fact, there is no stratification within the material, which therefore exhibits an isotropic monolithic behavior. Therefore, the existing calculation codes can be used.

### 4.4.2. *Printing using the method of injection into a particle bed*

The technology of printing through injection into a particle bed is also an additive manufacturing process, which is done layer-by-layer. Thus, a homogeneous distribution of the injected fluid (water in a bed of cement and aggregates or cement paste in a bed of aggregates) will directly influence the monolithic character of the structure and may or may not provide it with an anisotropic behavior.

Thus, for the technique of selective cement activation (injection of water into a bed of cement and aggregates), the heterogeneous distribution of the fluid phase leads to the heterogeneity of the material over its print height, which will provide the printed material with an anisotropic behavior. In addition, several studies have shown that the compressive strength of samples printed with this method increases with the water/cement mass ratio, which is contrary to the results that are expected and is completely counter-intuitive (normally, the strength increases when the water/cement ratio decreases, because there is less excess water for the hydration reaction, and therefore ultimately less porousness with the resultant damaging effects) [LOW 15, XIA 16, SHA 17]. In addition, the mechanical compressive strength obtained remains low, which is less than 20 MPa in all cases. These results can be explained by the absorption of water or fluids by the grains of the particle bed and by the lateral diffusion of the fluid phase, ultimately creating a heterogeneous level of water content within the material. As explained in the previous chapters, further work is needed to fully understand and control the infiltration of the fluid phase during injection.

Conversely, the technique of injecting cement paste into a bed of aggregates has made it possible to obtain compressive strengths of over 70 MPa (water/cement mass ratio of the cement paste W/C = 0.3, thickness of layers 3 mm), in accordance with the models predicting compressive strengths [WEG 18]. However, as shown by the initial studies carried out with this technique [PEG 97, WEG 16], a poor control of the depth of penetration of the cement paste (depending on its rheology), especially with an incomplete filling of the particle bed, will induce a loss of resistance and an anisotropic behavior of the layered material [PIE 18].

## 4.5. Conclusion

3D printing with concrete is different from conventional concrete casting in terms of the processes it uses for mixing and routing, the absence of formworks, etc.

Thus, the various stages of production (particularly layer-by-layer extrusion, injection or deposition) may lead to a stratified and/or orthotropic behavior, changing the behavior of a conventional isotropic and monolithic concrete.

As a result, the use of the standards and test methods used for conventional concrete may not be appropriate for printed materials and structures. It will be necessary to revise the criteria and the new regulations to measure and evaluate the mechanical performance of the 3D printing of concrete, as well as to develop new theoretical models to assess their structural behavior. New design standards and evaluation criteria are very important for ensuring the use of 3D printing elements and structures in order to be able to support all loads.

## 4.6. References

[AMZ 18] AMZIANE S., PERROT A., SONEBI M., "On some challenges to design printed formwork", *ETH Zurich, 1st International Conference on Concrete and Digital Fabrication*, September 9–12, 2018.

[AUS 12] AUSTIN S.A., LE T.T., LIM S. et al., "Hardened properties of high-performance printing concrete", *Cement Concrete Res*, vol. 42, pp. 558–566, 2012.

[FEN 15] FENG P., MENG X., CHEN J.F., et al., "Mechanical properties of structures 3D printed with cementitious powders", *Constr Build Mater*, vol. 93, pp. 486–497, 2015.

[GOS 16] GOSSELIN C., DUBALLET R., ROUX P. et al., "Large-scale 3D printing of ultra-high performance concrete—A new processing route for architects and builders", *Mater Des*, vol. 100, pp. 102–109, 2016.

[LE 12] LE T.T., AUSTIN S.A., LIM S. et al., "Microstructure of extruded cement-bonded fiber board", *Mater Struct*, vol. 45, pp. 1221–1232, 2012.

[LI 01] LI V.C., WANG S., WU C., "Tensile strain-hardening behavior of polyvinyl alcohol engineered cementitious composite (PVA-ECC)", *ACI Mater. J.-Am. Concr. Inst.*, vol. 98, pp. 483–492, 2001.

[LI 03] LI V.C., "On engineered cementitious composites (ECC)", *Journal of Advanced Concrete Technology*, vol. 1, pp. 215–230, 2003.

[LIM 12] LIM S., BUSWELL R.A., LE T.T. et al., Developments in construction-scale additive manufacturing processes, *Automation in Construction*, vol. 21, pp. 262–268, 2012.

[LLO 13] LLORET E., SHAHAB A.R., LINUS M. et al., "Complex concrete structures: Merging existing casting techniques with digital fabrication", *Mater. Ecol.*, vol. 60, pp. 40–49, March 2015.

[LOW 15] LOWKE D., WEGER D., HENKE K. et al., "3D-Drucken von Betonbauteilen Durch Selektives Binden Mit Calciumsilikatbasierten Zementen – Erste Ergebnisse Zu Beton-Technologischen Und Verfahrenstechnischen Einflüssen", *Tagungsbericht '19. Internationale Baustofftagung*, P. D.-I. H.-M. Ludwig, Weimar, 2015.

[MAL 15] MALAEB Z., HACHEM H., TOURBAH A. et al., "3D concrete printing: Machine and mix design", *International Journal of Civil Engineering*, vol. 6, pp. 14–22, 2015.

[MEL 13] MÉLINGE Y., HOANG V.H., RANGEARD D. et al., "Study of tribological behavior of fresh mortar against a rigid plane wall", *European Journal of Environmental and Civil Engineering*, vol. 17, pp. 419–429, 2013.

[NER 16] NERELLA V.N., "CON Print3D- 3D printing technology for onsite construction", *Concr Australia*, vol. 42, pp. 36–39, 2016.

[NER 17] NERELLA V.N., SCHROEFL C., YAZDI M.A. et al., "Micro-and macroscopic investigations of the interface between layers on the interface of 3D-printed cementitious elements", *Journal of Materials in Civil Engineering*, vol. 29, 2017.

[OGU 18] OGURA H., NERELLA V.N., MECHTCHERINE V., "Developing and testing of strain-hardening cement-based composites (SHCC) in the Context of 3D-Printing", *Materials (Basel)*, vol. 11, no. 8, August 2018.

[PEG 97] PEGNA J., "Exploratory investigation of solid freeform construction", *Automation in Construction*, vol. 5, pp. 427–437, 1997.

[PEL 03] PELED A., SHAH S.P., "Processing effects in cementitious composites: Extrusion and casting", *Journal of Materials Civil Engineering*, vol. 15, pp. 192–199, 2003.

[PER 06] PERROT A., Conditions d'extrudabilité des matériaux à base cimentaire, PhD Thesis, Rennes, INSA, 2006.

[PER 08] PERROT A., MOLEZ L., MELINGE Y. et al., "Influence des techniques de mise en forme sur les propriétés physiques d'un mortier extrudable, *GEODIM'08 "Variations Dimensionnelles Des Géomatériaux"*, Saint-Nazaire, France, 2008.

[PER 09] PERROT A., MÉLINGE Y., ESTELLÉ P. et al., "Vibro-extrusion: A new forming process for cement-based materials", *Advances in Cement Research*, vol. 21, pp. 125–133, 2009.

[PER 18] PERROT A., RANGEARD D., AMZIANE S. et al., "Optimizing fresh properties of concrete for extrusion and 3d printing", *RILEM 1st International Conference on Digital Fabrication with Concrete, Extended Abstracts*, pp. 45–46, September 9–12, 2018.

[PIE 18] PIERRE A., WEGER D., PERROT A. et al., "Penetration of cement pastes into sand packings during 3D printing: analytical and experimental study", *Materials and Structures*, vol. 51, no. 22, available at: https://doi.org/10.1617/s11527-018-1148-5, 2018.

[ROU 10] ROUSSEL N., CUSSIGH F., "Distinct-layer casting of SCC: The mechanical consequences of thixotropy", *Cement and Concrete Research*, vol. 38, pp. 624–632, 2010.

[RUB 17] RUBIO M., SONEBI M., AMZIANE S., "Fresh and rheological properties of 3D printing bio-cement-based materials", *Proceedings of 2nd ICBBM (PRO 119)*, Clermont-Ferrand, pp. 491–499, June 21–23, 2017.

[SAN 18] SANJAYAN J.G., NEMATOLLAHI B., XIA M. et al., "Effect of surface moisture on inter-layer strength of 3D printed concrete", *Construction and Building Materials*, vol. 172, pp. 468–475, 2018.

[SHA 01] SHAO Y., QIU J., SHAH S.P., "Microstructure of extruded cement-bonded fiber board", *Cement and Concrete Research*, vol. 31, pp. 1153–1161, 2001.

[SHA 17] SHAKOR P., SANJAYAN J., NAZARI A. et al., "Modified 3D printed powder to cement-based material and mechanical properties of cement scaffold used in 3D printing", *Construction and Building Materials*, vol. 138, pp. 398–409, 2017.

[SON 18] SONEBI M., RUBIO M., AMZIANE S. et al., "Mechanical properties of 3d printing bio-based fiber cement-based materials", *RILEM 1st International Conference on Digital Fabrication with Concrete, Extended Abstracts*, pp. 50–51, September 9-12, 2018.

[WAN 16] WANGLER T., LLORET E., REITER L. et al., "Digital concrete: Opportunities and challenges", *RILEM Technical Letters*, vol. 1, 2016DO - 1021809rilemtechlett201616, 2016.

[WEG 16] WEGER D., LOWKE D., GEHLEN C., "3D printing of concrete structures using the selective binding method–Effect of concrete technology on contour precision and compression strength", *Proceedings of 11th Fib International PhD Symposium in Civil Engineering*, The University of Tokyo, Tokyo, pp. 403–410, 2016.

[WEG 18] WEGER D., LOWKE D., GEHLEN C., "Additive manufacturing of concrete elements using the selective paste intrusion – effect of layer orientation on strength and durability", *Proceedings of RILEM 1st International Conference on Concrete and Digital Fabrication*, 2018.

[XIA 16] XIA M., SANJAYAN J., "Method of formulating geopolymer for 3D printing for construction applications", *Materials & Design*, vol. 110, pp. 382–390, 2016.

# 5

# 3D Printing with Concrete: Impact and Designs of Structures

## 5.1. Introduction

The introduction of digital control and the robotization of the construction industry has the potential to make a big change in the way buildings and civil engineering structures are designed and built.

This technology encourages us to reconsider the way in which buildings are designed, in order to have structures that optimize the forms and the quantities of materials used. Similarly, it is imperative to review the way in which the cement-based materials are reinforced, in order to compensate for their weakness under tensile stress, as well as to create standards and codes that design these new printed constructions well.

Similarly, "digital concrete" should shake up the organizational structure of a construction project and the skill sets required to build a structure, both during the design stage (engineering) and the production stage (construction). This societal impact of "raising skill levels" in the construction industry will have to be accompanied by a reduction in physical labor and the stress levels generated by work.

---

Chapter written by Arnaud PERROT and Damien RANGEARD.

Finally, the economic and environmental impact of concrete printing is generated both by the change of construction methods (suppression of formworks) and by the optimization of the use of available resources.

Thus, this chapter will build upon the previous chapters to assess and give an overview of the consequences and impacts of 3D printing on architecture, structural design, society, the economy and the environment.

## 5.2. Freedom of forms: architectural liberation and topological optimization

### 5.2.1. *3D printing with concrete: a boon for architects?*

The 3D printing of cement-based materials provides unprecedented freedom of form for architects and designers of concrete parts. The recent creation of artificial concrete reefs by the company XtreeE [XTR 18] shows the complexity of the shapes and textures that can be made through the additive manufacturing methods of cement-based materials (Figure 5.1).

**Figure 5.1.** *Example of complex structures made by the 3D printing of concrete: artificial underwater reefs. The numerous air holes created by the concrete encourage underwater fauna to settle in the area [XTR 18]. For a color version of the figures in this chapter see www.iste.co.uk/perrot/3dprinting.zip*

Thus, the additive manufacturing method applied to the concrete can produce parts for a single purpose without having to first manufacture expensive molds or formworks. From this point of view, this new tool may open up new fields of opportunities that were previously inaccessible to architects. The example of the Dfab House in Switzerland demonstrates the freedom that architects can achieve with these new techniques of digital construction, by designing structures with shapes that are both complex, and original, and optimally designed, as we will see in the following section (Figure 5.2) [DFA 18].

**Figure 5.2.** *The Dfab House: a building that demonstrates a concrete application of additive and digital manufacturing in the construction world [DFA 18]*

However, this method does not offer complete freedom of the forms that can be fabricated, especially when using extrusion/ deposition technology. Overhanging parts of structures are in fact limited by the elastic properties of the material while it is in a fresh state, as shown by Wolfs *et al.* [WOL 18] and in Chapter 3 of this book. One solution to this problem is the use or printing of temporary supports, as with the technique of polymer printing by the fused deposition modeling of materials. Printing supports made from recyclable materials appears to be a promising solution, making it possible to further extend the freedom that architects can obtain.

## 5.2.2. *Towards the creation of structures with optimized shapes?*

Digital production can also help to optimize the quantities of materials installed because the materials are only placed where they are necessary for the mechanical stability of the structure. Thus, the design of building and construction structures can be done using the concept of topological optimization. In many areas of application, additive manufacturing is related to a design using this topological optimization [BRA 11, HOL 05].

Topological optimization is a design tool that uses mathematical methods that allow the amounts of materials to be minimized in a given volume subjected to mechanical stress [BEN 01]. An illustration of the method on a beam subjected to a load in the middle of its span on two simple supports is shown in Figure 5.3, as detailed in Vantyghem *et al.* [VAN 18].

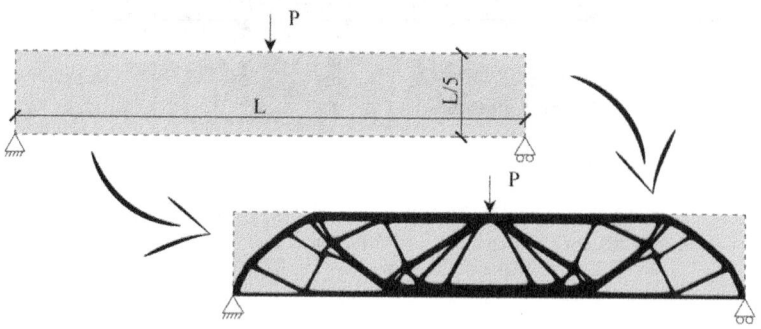

**Figure 5.3.** *Example of the application of topological optimization on a beam with two simple supports, with a load placed on the middle of the span [VAN 18]*

It is also possible to produce designs inspired by nature and the living world that, through a slow process of natural selection, has ultimately been able to produce structures that optimize the use of materials [SAN 05] to design load-bearing elements that can reduce the volume of the matter.

In this sense, a comparison between the porous structure of human bones and the column produced by the company XtreeE, Aix-en-Provence indicates that concrete 3D printing can be used to construct optimized structures inspired by nature (Figure 5.4) [ECH 16].

Another feature of digital concrete manufacturing is to build structures that are only subjected to compression stress, similar to masonry structures in the form of an arch or a dome. This method allows the mechanical weaknesses and the natural tensile fragility of the cement materials to be overcome. This strategy has been particularly used by the group of Pr. Block of ETH Zurich [RIP 13, VEE 14]. The idea is to use the techniques of the digital manufacturing of formworks based on "reinforcing" fabrics and cables as a support for the projection of a thin layer of concrete to be used in compression (Figure 5.5). Other digital production methods using concrete (injection into particle beds or by extrusion/deposition) may be used in the future in an attempt to reproduce traditional architectural forms, working in compression, that are found in cathedrals (arches and keystones without metal reinforcements).

**Figure 5.4.** *Support column with porous structures produced by the company XtreeE using digital production [XTR 18]*

**Figure 5.5.** *Structures made by minimizing tensile forces in concrete [EEV 14]*

### 5.2.3. *Could 3D concrete printers go through a transition similar to the transition from black and white to color?*

Many research projects today focus on simultaneous and collaborative works performed by a team of robots [ZHA 18, DES 18]. As with paper printers, a printing technique can use multiple inks: concretes of different densities and strengths, thermal insulators and reinforcements. Moreover, the resulting combinations are limitless and thus could be beneficial for using printing with different materials: a deposited thermal insulator could serve as concrete supports for complex geometries [DUB 18] (Figure 5.6), and as with the "mesh mold" technique, the printing of metal reinforcements could help retain the fresh concrete [HAC 15]. It should be noted that the full-scale printing of a polyurethane foam building has already been reported for the construction of the Yhnova house in Nantes [POU 18]. Thus, the printing of this material can be combined with that of concrete.

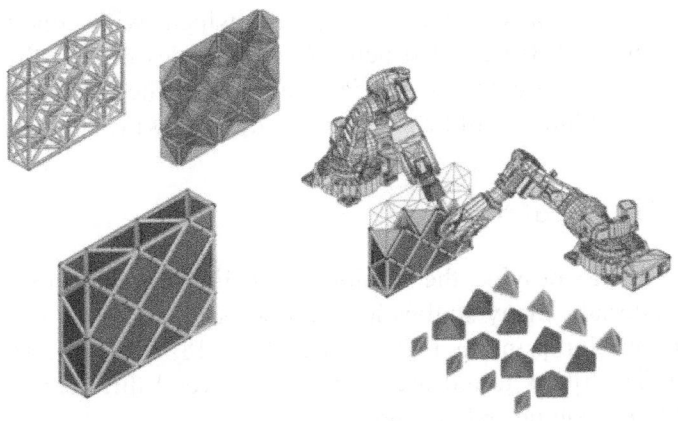

**Figure 5.6.** *Production of concrete walls (gray)/insulation (glued blocks) with two robots working together [DUB 18]*

Several research works have also dealt with the printing of concrete metal reinforcements [MEC 18a, MEC 18b, MEC 18c]. This could lead to the development of the printing of reinforced concrete. Meanwhile, this raises the question concerning the methods for reinforcing the printed concrete. This problem is addressed in the following section.

## 5.3. Design of structures: reinforcement strategies and design codes

While the structures developed by Philippe Block and his team are only subjected to compression and do not require reinforcements to compensate for the weakness of the concrete under tensile stress, many other printed concrete structures require the addition of reinforcements to ensure their mechanical stability. Several strategies have been tested, such as the addition of fibers [OGU 18, FAR 16], the addition of external reinforcements [ASP 18a], the installation of a steel wire inserted into the material during extrusion [BOS 18a] or the creation of a dedicated void for the later placement of post-tensioning cables or steel rebars [BOS 18b, LIM 12]. The development of reinforcement strategies for structures made from printed concrete is therefore a primary subject of study, which will condition the establishment of dedicated structural design codes [ASP 18b]. Before presenting the problem of standardization and the structural design codes, each solution will be presented in the following sections.

### 5.3.1. *The use of fibers*

In order to provide the minimum tensile properties for cement-based materials, many studies have proposed the addition of fibers to the mixture for printing [MEC 18c, OGU 18, PAN 17, HAM 17, LE 12]. The fibers tested are made from several different materials: glass, basalt, polypropylene and steel.

Collectively, these studies demonstrate an increase in the ductility of the material with a significant increase in tensile strength under flexural load. In their study, Ogura *et al.* [OGU 18] were able to formulate a strain-hardening material.

In addition to the mechanical contribution, the addition of fibers allows the creation of a fiber-reinforced cement material, thus making it possible to rely on validated design methods for these materials, such as the fib Model Code 2010 [BEV 13]. However, as mentioned in Chapter 4, the addition of fibers, due to their orientation induced by the extrusion process, makes the cement-based material highly anisotropic. Therefore, we must take into account the orientation of the fibers, and even the influence they have during extrusion and deposition using the theories based on the rheology of fiber-reinforced concrete [FER 15, MAR 10].

However, in their review of the literature on the reinforcement of printed concrete structures, Asprone *et al.* [ASP 18b] indicate that the addition of fibers may not be sufficient during high tensile stress. Moreover, the interface between deposits must be managed so that the fibers penetrate between the layers. Otherwise, the connections between layers would become mechanically weak under tension.

## 5.3.2. External reinforcements

Asprone *et al.* [ASP 18a] proposed the construction of a beam made up of segments assembled by external steel bars used as reinforcement. The bars are anchored in the segments using a sealing mortar, which enable the materials to have sufficient mechanical characteristics for bending (Figure 5.7).

This solution is satisfactory from the point of view of mechanical performance. However, the reinforcement material must be resistant to corrosion to allow an acceptable durability of the printed structure. In this case, the use of stainless steel or carbon, for example, is essential.

**Figure 5.7.** *Example of external reinforcements: (a) a 3D printed concrete segment; (b) connection details of external reinforcements made from steel; (c) beams with straight sections; and (d) variable sections obtained by mounting external reinforcements [26]*

## 5.3.3. Steel wire placed within the extruded material

An alternate solution was proposed by researchers at the Technical University of Eindhoven. It consists of the placement of a steel wire through the extrusion of the cement-based material. The continuous cable is contained in the layer of deposited concrete and provides bending strength to the printed cement material [BOS 18a, BOS 17].

However, the effect of the reinforcement is limited to the directions perpendicular to the deposition, and the diameter of the cable is reduced so that it is flexible during the printing. Hence, the effectiveness of this reinforcement technique is limited. These studies have used high-performance steel cables of diameters 0.63–1.2 mm, which differ from the standard diameters used in the field of reinforced concrete. In addition, problems involving sliding cables at the mortar interface have been reported, especially for cables with the highest mechanical strength (with no cable breakage during bending tests).

This method must therefore be made more reliable in order to successfully design structural components such as internal reinforcement cables. Thus, the design of the anchor points and the quality of the cable/matrix interface must be more clearly defined in order to arrive at a reliable and optimal structural design using this reinforcement technique.

### 5.3.4. Dedicated spaces acting as lost formworks

Another technique is to leave dedicated spaces for placing steel rebars later. The purpose of this method is to use the printed structure as a lost formwork, or as prefabricated elements that must be assembled or inserted using steel and/or concrete. Within this framework, the existing design codes can be approximated for the structural design of the structure. This use of printed structures is in keeping with conventional building methods, and thus allows the use of the Eurocodes standards.

In this case, several solutions are possible. The researchers from Eindhoven [BOS 18a] and Loughborough [LIM 09, LE 12] have provided some reservations about printed elements that must be assembled later using post-tensioning cables. For example, the use of post-tensioning technology has made it possible to build a bicycle bridge from printed segments (Figure 5.8).

**Figure 5.8.** *Bridge made from printed voussoirs and assembled using the post-tensioning technique – view of the anchor blocks [BOS 18a]*

Finally, a method for building walls using Contour Crafting [WU 16, KHO 04] is based on classical masonry walls. During the construction of these types of vertical walls, vertical reservations are made for the later implementation of a linked construction of classical masonry structures, which are commonly found in concrete block masonry used in individual houses. These methods are used by companies such as Apis Cor or WinSun. In this case, the printed structures play the role of lost formworks in which a conventional reinforced concrete is used by adding steel bars and pouring concrete. Thus, for this reinforcement technique, the Eurocodes and design codes related to reinforced concrete, as well as the design methods related to the conventional masonry structures formed from concrete blocks, may be applied.

## 5.3.5. *Wrapping of reinforcement elements set in place beforehand*

This solution was proposed by the company HuaSheng Tengda Ltd, in which two nozzles are used to deposit concrete, layer-by-layer, on either side of a cage of reinforcements that has previously been installed. This technique has been described in De Schutter *et al.* [DES 18].

This method is similar to that of the mesh mold technique, in which the concrete is cast in an optimized and printed steel mesh that acts as a reinforcement for a load-bearing wall with a complex geometry [HAC 15].

### 5.3.6. *Towards a specific design code?*

In order to justify the printed concrete structures mechanically, a dedicated structural design code is necessary to take into account the specific elements of the printed materials. As a result, a classification of the different printing techniques and reinforcement strategies (which can lead to the anisotropy of the material) is necessary. Meanwhile, the use of reinforcements placed subsequently in dedicated spaces would allow existing standards to be used, including Eurocodes 2 and the use of fibers on the fib Model Code 2010.

## 5.4. Impacts of 3D printing

As noted earlier in this chapter, the application of 3D printing in the field of concrete construction could lead to a big change not only in the way in which construction project is carried out, but also in their design. The changes that the introduction of 3D printing will bring to the construction sector should have a significant impact on both the socio-economic activity of the construction sector and its environmental impact. In this section, we will attempt to address these impacts.

### 5.4.1. *Environmental impact*

The integration of 3D printing with methods for concrete construction will have an environmental impact at different levels of the building process.

First, additive manufacturing should limit or even eliminate the production and management of formworks, which can generate a large amount of waste, particularly in the case of formworks for complex structures with assembly components that are used only once. However, it should be noted that the impact of eliminating this management of formworks will have a rather marginal effect on the environmental impact of construction. It is generally considered that carbon emissions due to the on-site construction period account for only 2% of the overall emissions related to a construction work during its life cycle [DES 18, HON 15]. It is also important to note that the manufacturing of the tools required for 3D printing (robotic arms, for example) will only have a negligible effect on the environmental impact of the construction.

In this sense, Agusti-Juan and Habert [AGU 17] indicated that the reduction in environmental impacts related to 3D printing mainly comes from the design phase, which allows the topological optimization of structures and thus reduces the volumes of materials involved in a given mechanical function. An example is given in Figure 5.9, showing a simple case of a rectangular wall [DES 18]. This feature offers the advantage of a potential reduction in the supplies of raw materials, especially aggregates, which have become increasingly scarce.

Gains could also be made through the multiple functions of printed materials: for instance, the optimization of concrete forms can boost thermal and acoustic performances, and thus reduce the thickness of insulation materials. In addition, the shape of the parts can be used to set up electric, air or water supply networks.

However, it is important to note that the formulas of cement-based materials tested in the literature often use reduced maximum grain sizes, which significantly increases the cement dosage. Similarly, a significant use of chemical admixtures has also been reported in many cases. As a result, the environmental impact of the printable material is greater than that of conventional concrete. Thus, for the printed concrete to have less of an impact, the design must be optimized, in order to, at a minimum, compensate for the higher environmental cost of the formulas used for printing.

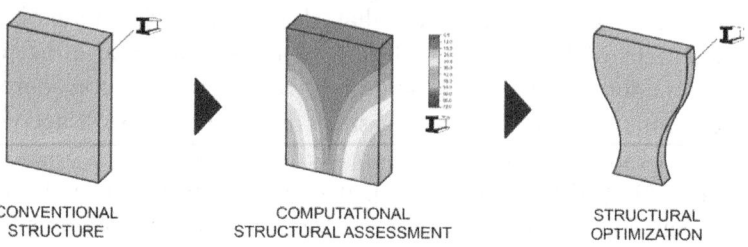

CONVENTIONAL STRUCTURE    COMPUTATIONAL STRUCTURAL ASSESSMENT    STRUCTURAL OPTIMIZATION

**Figure 5.9.** *Visualization of the reduction in quantities of materials using topological optimization during the design phase*

### 5.4.2. Societal impact

The introduction of 3D printing in the construction industry could have repercussions in terms of the involved work and the organization of construction sites. The elimination of the steps involving formworks (assembly, casting, disassembly) reduces the number of manual tasks and the handling of heavy materials.

This has led to safety improvements and an increase in the technical level of construction work, thereby increasing the number of computer-assisted manufacturing managers and skilled robotic technicians on building sites.

The view generally held on the industrial sector of building and construction, which suffers from an image of being "low tech" as opposed to "high tech", is likely to be substantially improved. In fact, it is worth noting that the construction industry suffers from an image of being an old practice, with productivity levels that are at best stagnant and a limited willingness to invest in research and development. The introduction of digital construction could "finally" change these perceptions, promote truly dynamic research into the methods of carrying out construction projects and have a positive influence on the state of mind of construction companies regarding innovation.

### 5.4.3. *Economic impact*

From an economic point of view, 3D printing will speed up production rates and thus improve productivity in the sector. Gains in quality levels through the use of digital production tools developed for building design could also lead to an increase in corporate profit levels (or a reduction in costs).

In terms of construction materials, the elimination of the casings and stages required for concrete castings, which represent 30–50% of the cost of a construction site depending on its design and construction [DES 18, CHE 02], should lead to a significant reduction in construction costs. These savings will be greater than the cost of the new printing equipment in order for the removal of casings to be viable. De Schutter *et al.* [DES 18] explain that the cost of printers is closely linked to the use (and repurposing) of conventional construction equipment such as pumps, cranes and crane trucks.

In addition, labor costs are expected to decrease, while the skill levels required to manage and maintain the robotic equipment for construction should increase.

At the material level, even though topological optimization saves concrete, printable concrete contains many relatively expensive additives (especially those intended to "boost" the speed of mechanical structuring) that will compensate for the gains in the reduction of volumes.

Finally, the ways in which 3D printing may become economically competitive can be found in all its different aspects (methods, equipment, labor and materials). Each of these possible paths presents potential gains, but also additional costs that are inherent to the process and must be quantified and accounted for.

A recent study on the construction of the NEST in Switzerland showed that for simple geometries, conventional construction was less expensive than digital construction. However, when the geometry became more complex, digital construction began to increase savings [DE 18].

It is important to note that at that time, the process of 3D printing with concrete, and that of digital construction in general, was in its infancy. As a result, progress must be made in many areas, which will undoubtedly make it increasingly competitive and thus more attractive in the future. Thus, the economic impact of concrete 3D printing must be reassessed later.

## 5.5. Conclusion

This chapter has shown the big change that digital construction may bring about in the field of construction, both in terms of the design strategy of formworks and the level of the stage at which they are carried out.

The new-found freedom of forms allows the design of structures to follow the intrinsic qualities of concrete (reduction of zones subjected to tension) and the reduction of the amounts of materials used. In addition, innovative reinforcement techniques will be tested and developed to work together with the printing process. Currently, research is carried out on the simultaneous printing of concrete and steel reinforcements.

In the long-term, the use of printing in the construction sector should lead to a reduction in the environmental impact of structures (reduced volumes of materials, multiple functionalities of structures, etc.), and a reduction in construction costs through an increase in speed and quality. Finally, at the societal level, the construction industry will shift away from its "low tech" image to the one that seeks out innovation and an increase in skill levels needed to manage a construction site.

## 5.6. References

[AGU 17] AGUSTÍ-JUAN I., HABERT G., "Environmental design guidelines for digital fabrication", *Journal of Cleaner Production*, vol. 142, pp. 2780–2791, 2017.

[ASP 18a] ASPRONE D., AURICCHIO F., MENNA C. *et al.*, "3D printing of reinforced concrete elements: Technology and design approach", *Construction and Building Materials*, vol. 165, pp. 218–231, March 2018.

[ASP 18b] ASPRONE D., MENNA C., BOS F.P. *et al.*, "Rethinking reinforcement for digital fabrication with concrete", *Cement and Concrete Research*, 2018.

[BEN 01] BENDSØE M.P., "Topology optimization", in *Encyclopedia of Optimization*, Springer, pp. 2636–2638, 2001.

[BEV 13] BEVERLY P., *fib Model Code for Concrete Structures 2010*, Ernst & Sohn, 2013.

[BOS 17] BOS F.P., AHMED Z.Y., JUTINOV E.R. *et al.*, "Experimental exploration of metal cable as reinforcement in 3D printed concrete", *Materials*, vol. 10, no. 11, p. 1314, 2017.

[BOS 18a] BOS F.P., AHMED Z.Y., WOLFS R.J.M. *et al.*, "3D printing concrete with reinforcement", in *High Tech Concrete: Where Technology and Engineering Meet*, pp. 2484–2493, 2018.

[BOS 18b] BOS F., WOLFS R., AHMED Z. *et al.*, "Large scale testing of digitally fabricated concrete (DFC) elements", in *RILEM International Conference on Concrete and Digital Fabrication*, 2018.

[BRA 11] BRACKETT D., ASHCROFT I., HAGUE R., "Topology optimization for additive manufacturing", in *Proceedings of the Solid Freeform Fabrication Symposium*, Austin, TX, vol. 1, pp. 348–362, 2011.

[CHE 02] CHEN W.-F., LIEW J. R., *The Civil Engineering Handbook*, Crc Press, 2002.

[DE 18] DE SOTO B. G. *et al.*, "Productivity of digital fabrication in construction: Cost and time analysis of a robotically built wall", *Automation in Construction*, vol. 92, pp. 297–311, 2018.

[DES 18] DE SCHUTTER G., LESAGE K., MECHTCHERINE V. *et al.*, "Vision of 3D printing with concrete — Technical, economic and environmental potentials", *SI Digit. Concr. 2018*, vol. 112, pp. 25–36, October 2018.

[DFA 18] "DFAB HOUSE – Building with robots and 3D printers", Available at: http://dfabhouse.ch/, 2018.

[DUB 18] DUBALLET R., BAVEREL O., DIRRENBERGER J., *Design of Space Truss Based Insulating Walls for Robotic Fabrication in Concrete*, 2018.

[ECH 16] LES ECHOS, "L'impression 3D veut imiter la nature". Available at: https://www.lesechos.fr/13/12/2016/LesEchos/22338-046-ECH_l-impression-3d-veut-imiter-la-nature.htm, December 2016.

[FAR 16] FARINA I. *et al.*, "On the reinforcement of cement mortars through 3D printed polymeric and metallic fibers", *Composites Part B: Engineering*, vol. 90, pp. 76–85, 2016.

[FER 15] FÉREC J., PERROT A., AUSIAS G., "Toward modeling anisotropic yield stress and consistency induced by fiber in fiber-reinforced viscoplastic fluids", *Journal of Non-Newtonian Fluid Mechanics*, vol. 220, pp. 69–76, 2015.

[HAC 15] HACK N., LAUER W., GRAMAZIO F. *et al.*, "Mesh Mould: Robotically fabricated metal meshes as concrete formwork and reinforcement", in *Proceedings of the 11th International Symposium on Ferrocement and 3rd ICTRC International Conference on Textile Reinforced Concrete, Aachen, Germany*, pp. 7–10, 2015.

[HAM 17] HAMBACH M., VOLKMER D., "Properties of 3D-printed fiber-reinforced Portland cement paste", *Cement and Concrete Composites*, vol. 79, pp. 62–70, May 2017.

[HOL 05] HOLLISTER S.J., "Porous scaffold design for tissue engineering", *Nature Materials*, vol. 4, no. 7, p. 518, 2005.

[HON 15] HONG J., SHEN G.Q., FENG Y. *et al.*, "Greenhouse gas emissions during the construction phase of a building: a case study in China", *Journal of Cleaner Production*, vol. 103, p. 249–259, 2015.

[KHO 04] KHOSHNEVIS B., "Automated construction by contour crafting–related robotics and information technologies", *Best ISARC 2002*, vol. 13, no. 1, pp. 5–19, January 2004.

[LE 12] LE T.T. *et al.*, "Hardened properties of high-performance printing concrete", *Cement and Concrete Research*, vol. 42, no. 3, pp. 558–566, 2012.

[LIM 09] LIM S. et al., "Fabricating construction components using layered manufacturing technology", in *Global Innovation in Construction Conference*, pp. 512–520, 2009.

[LIM 12] LIM S., BUSWELL R.A., LE T.T. et al., "Developments in construction-scale additive manufacturing processes", *Automation in Construction*, vol. 21, pp. 262–268, January 2012.

[MAR 10] MARTINIE L., ROSSI P., ROUSSEL N., "Rheology of fiber reinforced cementitious materials: classification and prediction", *Cement and Concrete Research*, vol. 40, no. 2, pp. 226–234, 2010.

[MEC 18a] MECHTCHERINE V., GRAFE J., NERELLA V.N. et al., "3D-printed steel reinforcement for digital concrete construction–Manufacture, mechanical properties and bond behaviour", *Construction and Building Materials*, vol. 179, pp. 125–137, 2018.

[MEC 18b] MECHTCHERINE V., NERELLA V.N., "Incorporating reinforcement in 3D-printing with concrete", *Beton- Stahlbetonbau*, vol. 113, no. 7, pp. 496–504, 2018.

[MEC 18c] MECHTCHERINE V. et al., "Alternative reinforcements for digital concrete construction", in *RILEM International Conference on Concrete and Digital Fabrication*, pp. 167–175, 2018.

[OGU 18] OGURA H., NERELLA V.N., MECHTCHERINE V., "Developing and Testing of Strain-Hardening Cement-Based Composites (SHCC) in the Context of 3D-Printing", *Materials (Basel)*, vol. 11, no. 8, August 2018.

[PAN 17] PANDA B., PAUL S.C., TAN M.J., "Anisotropic mechanical performance of 3D printed fiber reinforced sustainable construction material", *Materials Letters*, vol. 209, pp. 146–149, 2017.

[POU 18] POULLAIN P., PAQUET E., GARNIER S. et al., "On site deployment of 3D printing for the building construction – The case of YhnovaTM", *MATEC Web Conference*, vol. 163, pp. 01001, 2018.

[RIP 13] RIPPMANN M., BLOCK P., "Rethinking structural masonry: Unreinforced, stone-cut shells", *Proceedings of the Institution of Civil Engineers - Construction Materials*, vol. 166, no. 6, pp. 378–389, December 2013.

[SAN 05] SANCHEZ C., ARRIBART H., GUILLE M.M.G., "Biomimetism and bioinspiration as tools for the design of innovative materials and systems", *Nature Materials*, vol. 4, no. 4, p. 277, 2005.

[VAN 18] VANTYGHEM G., BOEL V., DE CORTE W. et al., "Compliance, Stress-based and Multi-physics topology optimization for 3D-Printed concrete structures", in *RILEM International Conference on Concrete and Digital Fabrication*, pp. 323–332, 2018.

[VEE 14] VEENENDAAL D., BLOCK P., "Design process for prototype concrete shells using a hybrid cable-net and fabric formwork", *Engineering Structures*, vol. 75, pp. 39–50, 2014.

[WOL 18] WOLFS R.J.M., BOS F.P., SALET T.A.M., "Early age mechanical behaviour of 3D printed concrete: Numerical modelling and experimental testing", *Cement and Concrete Research*, vol. 106, pp. 103–116, April 2018.

[WU 16] WU P., WANG J., WANG X., "A critical review of the use of 3-D printing in the construction industry", *Automation in Construction*, vol. 68, pp. 21–31, August 2016.

[XTR 18] "Project – Rexcor Artificial Reef | XtreeE", available at: http://www.xtreee.eu/, 2018.

[ZHA 18] ZHANG X. et al., "Large-scale 3D printing by a team of mobile robots", *Automation in Construction*, vol. 95, pp. 98–106, November 2018.

# List of Authors

Sofiane AMZIANE
Institut Pascal
University of Clermont
Auvergne
France

Arnaud PERROT
IRDL
University of Southern
Brittany
Lorient
France

Alexandre PIERRE
L2MGC
Cergy-Pontoise University
France

Damien RANGEARD
LGCGM
INSA Rennes
France

Mohammed SONEBI
Queen's University Belfast
United Kingdom

# Index

**A, B, C**

anisotropy, 102, 105, 107, 109, 114, 115, 136
bending strength, 104, 109, 114, 134
bonding, 6, 65, 73–76, 79, 81, 84, 85, 103, 104, 114, 116
cement-based materials, 7–9, 11–14, 21, 22, 29, 41, 46–48, 54, 67, 73, 85, 101, 103, 105, 109, 114–116, 119, 125, 126, 132, 137
compressive strength, 106, 109, 110, 112, 114, 115, 117–120

**D, E, F, I, P**

design, 2, 3, 12, 20, 62, 74, 75, 120, 125, 126, 128, 131, 132, 134–140
digital
 construction, 25, 41, 127, 138, 140
 production, 128, 129, 139
extrusion, 9, 41, 54, 62, 102, 103

fibers, 21, 64, 83, 104, 105, 109, 110, 112, 113, 115, 116, 131, 132, 136
freedom of architects, 21
impact
 economic, 139, 140
 environmental, 126, 136, 137, 140
 societal, 125, 138
interface, 21, 44, 54, 65, 66, 68, 79, 104, 107, 109, 111, 114–116, 132, 134
penetration, 80, 81, 84, 85, 87, 89–96, 120
permeability, 45, 65
precision, 22, 31, 74, 75
pumping, 9, 10, 41–43, 45–47, 49, 50, 52–54

**R, S, T, V**

reinforcement, 21, 76, 131–136, 140
reinforcing, 129, 131

rheology, 9, 25–27, 30, 43, 46, 49, 53, 81, 90, 93, 106, 120, 132
selective intrusion, 80, 82, 83, 89, 90, 92, 94–96
shear yield stress, 45, 47, 49–54, 57, 60, 62, 96, 103, 116, 117
shrinkage, 63–65
slip-forming, 8, 9, 25
stereolithography (SLA), 4
stereolithography format (STL), 3, 74
structural build-up rate, 15, 45, 50, 60
thixotropy, 92, 96, 116
topological optimization, 83, 126, 128, 137–139
viscosity, 45, 48, 53, 79, 85, 86, 90, 107

Other titles from

in

Civil Engineering and Geomechanics

## 2019

LAMBERT David Edward, PASILIAO Crystal L., ERZAR Benjamin, REVIL-BAUDARD Benoit, CAZACU Oana
*Dynamic Damage and Fragmentation*

## 2018

FROSSARD Etienne
*Granular Geomaterials Dissipative Mechanics: Theory and Applications in Civil Engineering*

KHALFALLAH Salah
*Structural Analysis 1: Statically Determinate Structures*
*Structural Analysis 2: Statically Indeterminate Structures*

VERBRUGGE Jean-Claude, SCHROEDER Christian
*Geotechnical Correlations for Soils and Rocks*

## 2017

LAUZIN Xavier
*Civil Engineering Structures According to the Eurocodes*

PUECH Alain, GARNIER Jacques
*Design of Piles Under Cyclic Loading: SOLCYP Recommendations*

SELLIER Alain, GRIMAL Étienne, MULTON Stéphane, BOURDAROT Éric
*Swelling Concrete in Dams and Hydraulic Structures: DSC 2017*

## 2016

BARRE Francis, BISCH Philippe, CHAUVEL Danièle *et al.*
*Control of Cracking in Reinforced Concrete Structures: Research Project CEOS.fr*

PIJAUDIER-CABOT Gilles, LA BORDERIE Christian, REESS Thierry, CHEN Wen, MAUREL Olivier, REY-BETHBEDER Franck, DE FERRON Antoine
*Electrohydraulic Fracturing of Rocks*

TORRENTI Jean-Michel, LA TORRE Francesca
*Materials and Infrastructures 1*
*Materials and Infrastructures 2*

## 2015

AÏT-MOKHTAR Abdelkarim, MILLET Olivier
*Structure Design and Degradation Mechanisms in Coastal Environments*

MONNET Jacques
*In Situ Tests in Geotechnical Engineering*

## 2014

DAÏAN Jean-François
*Equilibrium and Transfer in Porous Media – 3-volume series*
*Equilibrium States – Volume 1*
*Transfer Laws – Volume 2*
*Applications, Isothermal Transport, Coupled Transfers – Volume 3*

## 2013

AMZIANE Sofiane, ARNAUD Laurent
*Bio-aggregate-based Building Materials: Applications to Hemp Concretes*

BONELLI Stéphane
*Erosion in Geomechanics Applied to Dams and Levees*

CASANDJIAN Charles, CHALLAMEL Noël, LANOS Christophe,
HELLESLAND Jostein
*Reinforced Concrete Beams, Columns and Frames: Mechanics and Design*

GUÉGUEN Philippe
*Seismic Vulnerability of Structures*

HELLESLAND Jostein, CHALLAMEL Noël, CASANDJIAN Charles,
LANOS Christophe
*Reinforced Concrete Beams, Columns and Frames: Section and Slender Member Analysis*

LALOUI Lyesse, DI DONNA Alice
*Energy Geostructures: Innovation in Underground Engineering*

LEGCHENKO Anatoly
*Magnetic Resonance Imaging for Groundwater*

# 2012

BONELLI Stéphane
*Erosion of Geomaterials*

JACOB Bernard *et al.*
*ICWIM6 – Proceedings of the International Conference on Weigh-In-Motion*

OLLIVIER Jean-Pierre, TORRENTI Jean-Marc, CARCASSES Myriam
*Physical Properties of Concrete and Concrete Constituents*

PIJAUDIER-CABOT Gilles, PEREIRA Jean-Michel
*Geomechanics in $CO_2$ Storage Facilities*

# 2011

BAROTH Julien, BREYSSE Denys, SCHOEFS Franck
*Construction Reliability: Safety, Variability and Sustainability*

CREMONA Christian
*Structural Performance: Probability-based Assessment*

HICHER Pierre-Yves
*Multiscales Geomechanics: From Soil to Engineering Projects*

IONESCU Ioan R. *et al.*
*Plasticity of Crystalline Materials: from Dislocations to Continuum*

LOUKILI Ahmed
*Self Compacting Concrete*

MOUTON Yves
*Organic Materials for Sustainable Construction*

NICOT François, LAMBERT Stéphane
*Rockfall Engineering*

PENSÉ-LHÉRITIER Anne-Marie
*Formulation*

PIJAUDIER-CABOT Gilles, DUFOUR Frédéric
*Damage Mechanics of Cementitious Materials and Structures*

RADJAI Farhang, DUBOIS Frédéric
*Discrete-element Modeling of Granular Materials*

RESPLENDINO Jacques, TOUTLEMONDE François
*Designing and Building with UHPFRC*

# 2010

ALSHIBLI A. Khalid
*Advances in Computed Tomography for Geomechanics*

BUZAUD Eric, IONESCU Ioan R., VOYIADJIS Georges
*Materials under Extreme Loadings / Application to Penetration and Impact*

LALOUI Lyesse
*Mechanics of Unsaturated Geomechanics*

NOVA Roberto
*Soil Mechanics*

SCHREFLER Bernard, DELAGE Pierre
*Environmental Geomechanics*

TORRENTI Jean-Michel, REYNOUARD Jean-Marie, PIJAUDIER-CABOT Gilles
*Mechanical Behavior of Concrete*

## 2009

AURIAULT Jean-Louis, BOUTIN Claude, GEINDREAU Christian
*Homogenization of Coupled Phenomena in Heterogenous Media*

CAMBOU Bernard, JEAN Michel, RADJAI Fahrang
*Micromechanics of Granular Materials*

MAZARS Jacky, MILLARD Alain
*Dynamic Behavior of Concrete and Seismic Engineering*

NICOT François, WAN Richard
*Micromechanics of Failure in Granular Geomechanics*

## 2008

BETBEDER-MATIBET Jacques
*Seismic Engineering*

CAZACU Oana
*Multiscale Modeling of Heterogenous Materials*

HICHER Pierre-Yves, SHAO Jian-Fu
*Soil and Rock Elastoplasticity*

JACOB Bernard *et al.*
*HVTT 10*

JACOB Bernard *et al.*
*ICWIM 5*

SHAO Jian-Fu, BURLION Nicolas
*GeoProc2008*

## 2006

BALAGEAS Daniel, FRITZEN Claus-Peter, GÜEMES Alfredo
Structural Health Monitoring

DESRUES Jacques *et al.*
*Advances in X-ray Tomography for Geomaterials*

FSTT
*Microtunneling and Horizontal Drilling*

MOUTON Yves
*Organic Materials in Civil Engineering*

## 2005

PIJAUDIER-CABOT Gilles, GÉRARD Bruno, ACKER Paul
*Creep Shrinkage and Durability of Concrete and Concrete Structures
CONCREEP – 7*

Printed and bound by CPI Group (UK) Ltd, Croydon, CR0 4YY